APPROACHES TO AND PREPARATION FOR THE OPERATION OF SMALL MODULAR REACTORS

The following States are Members of the International Atomic Energy Agency:

AFGHANISTAN	GEORGIA	PAKISTAN
ALBANIA	GERMANY	PALAU
ALGERIA	GHANA	PANAMA
ANGOLA	GREECE	PAPUA NEW GUINEA
ANTIGUA AND BARBUDA	GRENADA	PARAGUAY
ARGENTINA	GUATEMALA	PERU
ARMENIA	GUINEA	PHILIPPINES
AUSTRALIA	GUYANA	POLAND
AUSTRIA	HAITI	PORTUGAL
AZERBAIJAN	HOLY SEE	QATAR
BAHAMAS, THE	HONDURAS	REPUBLIC OF MOLDOVA
BAHRAIN	HUNGARY	ROMANIA
BANGLADESH	ICELAND	RUSSIAN FEDERATION
BARBADOS	INDIA	RWANDA
BELARUS	INDONESIA	SAINT KITTS AND NEVIS
BELGIUM	IRAN, ISLAMIC REPUBLIC OF	SAINT LUCIA
BELIZE	IRAQ	SAINT VINCENT AND
BENIN	IRELAND	THE GRENADINES
BOLIVIA, PLURINATIONAL	ISRAEL	SAMOA
STATE OF	ITALY	SAN MARINO
BOSNIA AND HERZEGOVINA	JAMAICA	SAUDI ARABIA
BOTSWANA	JAPAN	SENEGAL
BRAZIL	JORDAN	SERBIA
BRUNEI DARUSSALAM	KAZAKHSTAN	SEYCHELLES
BULGARIA	KENYA	SIERRA LEONE
BURKINA FASO	KOREA, REPUBLIC OF	SINGAPORE
BURUNDI	KUWAIT	SLOVAKIA
CABO VERDE	KYRGYZSTAN	SLOVENIA
CAMBODIA	LAO PEOPLE'S DEMOCRATIC	SOMALIA
CAMEROON	REPUBLIC	SOUTH AFRICA
CANADA	LATVIA	SPAIN
CENTRAL AFRICAN	LEBANON	SRI LANKA
REPUBLIC	LESOTHO	SUDAN
CHAD	LIBERIA	SWEDEN
CHILE	LIBYA	SWITZERLAND
CHINA	LIECHTENSTEIN	SYRIAN ARAB REPUBLIC
COLOMBIA	LITHUANIA	TAJIKISTAN
COMOROS	LUXEMBOURG	THAILAND
CONGO	MADAGASCAR	TOGO
COOK ISLANDS	MALAWI	TONGA
COSTA RICA	MALAYSIA	TRINIDAD AND TOBAGO
CÔTE D'IVOIRE	MALI	TUNISIA
CROATIA	MALTA	TÜRKİYE
CUBA	MARSHALL ISLANDS	TURKMENISTAN
CYPRUS	MAURITANIA	UGANDA
CZECH REPUBLIC	MAURITIUS	UKRAINE
DEMOCRATIC REPUBLIC	MEXICO	UNITED ARAB EMIRATES
OF THE CONGO	MONACO	UNITED KINGDOM OF
DENMARK	MONGOLIA	GREAT BRITAIN AND
DJIBOUTI	MONTENEGRO	NORTHERN IRELAND
DOMINICA	MOROCCO	UNITED REPUBLIC OF TANZANIA
DOMINICAN REPUBLIC	MOZAMBIQUE	UNITED STATES OF AMERICA
ECUADOR	MYANMAR	URUGUAY
EGYPT	NAMIBIA	UZBEKISTAN
EL SALVADOR	NEPAL	VANUATU
ERITREA	NETHERLANDS,	VENEZUELA, BOLIVARIAN
ESTONIA	KINGDOM OF THE	REPUBLIC OF
ESWATINI	NEW ZEALAND	VIET NAM
ETHIOPIA	NICARAGUA	YEMEN
FIJI	NIGER	ZAMBIA
FINLAND	NIGERIA	ZIMBABWE
FRANCE	NORTH MACEDONIA	
GABON	NORWAY	
GAMBIA, THE	OMAN	

IAEA-TECDOC-2110

APPROACHES TO AND PREPARATION FOR THE OPERATION OF SMALL MODULAR REACTORS

INTERNATIONAL ATOMIC ENERGY AGENCY
VIENNA, 2025

For further information on this publication, please contact:

Nuclear Power Engineering Section
International Atomic Energy Agency
Vienna International Centre
PO Box 100
1400 Vienna, Austria
Email: Official.Mail@iaea.org

© IAEA, 2025
Printed by the IAEA in Austria
December 2025
https://doi.org/10.61092/iaea.hhnl-1026

IAEA Library Cataloguing in Publication Data

Names: International Atomic Energy Agency.
Title: Approaches to and preparation for the operation of small modular reactors / International Atomic Energy Agency.
Description: Vienna : International Atomic Energy Agency, 2025. | Series: IAEA TECDOC ISSN 1011-4289 ; no. 2110 | Includes bibliographical references.
Identifiers: IAEAL 25-01803 | ISBN 978-92-0-127825-8 (paperback : alk. paper) | ISBN 978-92-0-127725-1 (pdf)
Subjects: LCSH: Nuclear reactors — Design and construction. | Nuclear reactors — Safety measures. | Nuclear reactors — Maintenance and repair.

FOREWORD

Small modular reactors (SMRs) are advanced nuclear reactors, typically designed to generate up to 300 MW of electricity, that feature modularity to optimize construction schedules and reduce costs. SMRs are designed to create synergies between nuclear and distributed energy resources, particularly renewables, and subsequently tighten the coupling of these complementary technologies through various non-electrical applications. Together, these technologies are expected to contribute to climate change mitigation, for example by supporting the achievement of net zero emissions by 2050.

In 2025, more than 70 design concepts from major lines of technologies are at various stages of development and deployment. At least 20 Member States are engaged in national and international activities involving SMRs, and the commercial operation of SMRs is now under way. The first SMRs, of the water cooled reactor type, have been in commercial operation since May 2020 at a floating nuclear power plant (FNPP) in the Russian Federation since May 2020 generating 70 MW(e). A high temperature gas cooled SMR entered commercial operation in December 2023 in China, producing 200 MW(e). Construction projects for SMRs are in progress or planned to begin by the early 2030s in several Member States, including Canada, China, France, the Republic of Korea, Romania, the Russian Federation, the United Kingdom, the United States of America and Uzbekistan.

The IAEA and other international organizations have published a wide range of resources on SMRs, addressing topics such as design, technology, engineering, economics, safety, safeguards, security and infrastructure development. However, operational readiness for SMRs has not yet been specifically covered by existing publications. Over the course of various IAEA meetings and forums, Member States interested in SMRs have identified a need for up-to-date information about, and examples of practical experience with, operational readiness and safety issues prior to the commercial deployment of SMRs.

To address the identified gaps in technical publications for Member States, the IAEA initiated a dedicated collaboration among its internal experts in various fields, including nuclear power engineering and nuclear power technology development. As part of this collaboration, a task force was formed to develop the present publication, informed by a series of expert meetings, which consolidates information to help relevant stakeholders understand and prepare to support the commissioning and operation of SMRs.

This publication presents state of the art information on commissioning, the approach to criticality, connection to the grid, operational limits and conditions, load following operation, assurance of fuel and component integrity, main control room arrangement, human–machine interface and management of operating personnel.

The IAEA is grateful to the many experts from Member States who contributed to this publication. The IAEA officers responsible for this publication were J.H. Jung, W.B. Kim and M.H. Subki of the Division of Nuclear Power.

CONTENTS

1. INTRODUCTION

1.1. BACKGROUND

SMRs are advanced nuclear reactors that are designed to typically generate up to 300 MW(e) per unit. Modularity is applied in the design of structures, systems, and components (SSCs) for SMRs to achieve simplification, enhanced transportability, thus shorten construction time and reduce cost. There are over 70 designs of SMRs under different phases of developments from major lines of technology: (i) water cooled reactors, (ii) non-water cooled reactors that include high temperature gas cooled reactors, liquid metal cooled reactors with fast neutron spectrum, the coolants could be sodium, lead and lead-bismuth, and the molten salt reactors. In terms of siting or plant arrangement, SMRs can be deployed as land based or marine based nuclear power plants such as floating power unit located on shore or offshore. In the past few years, microreactor designs have emerged as a subcategory of SMRs. They are designed to generate electricity between 1 and 20 MW(e) and can be of one or more technology and siting categories mentioned above. Member States interest in SMRs is increasing. Significant institutional and industrial efforts are being devoted to facilitating their development and early deployment.

Deployment of SMRs have materialized, as in mid-2025 two nuclear power plants with SMR were in operation. The Akademik Lomonosov FNPP is in operation in Russian Federation generating 70 MW(e). In China, the High Temperature Reactor Pebble-bed Module (HTR-PM) is in commercial operation as an industrial demonstration high temperature gas cooled reactor, generating 200 MW(e). Several other SMR designs for near term deployment are also reported in this TECDOC. The ACP100 is a 125 MW(e) integral Pressurized Water Reactor (PWR) design under construction in China with target connection to the grid by end of 2026. The BWRX-300 is a natural circulation boiling water reactor (BWR) from the United States and Japan that has received license for construction in April 2025 for deployment in the Darlington nuclear power plant site, Canada to generate 300 MW(e) by 2030. In Argentina, CAREM25 is under construction. Rolls-Royce SMR is a 3 loop PWR undertaking licensing in the United Kingdom. System-integrated Modular Advanced ReacTor 100 (SMART100) is a PWR design from Republic of Korea. VOYGR™, a natural circulation PWR designed to generate 462 MW(e) from six (6) reactor modules has received standard design approval (SDA) in the United States.

SMRs are designed to generate baseload and flexible power to meet the expectation of forming synergistic mix of nuclear with distributed energy resources, particularly the renewables, as well as various non-electric applications. The implications of the mentioned power generation schemes of SMRs needs to be identified and understood in terms of the OLCs, reliable and safe operation, economics, fuel management and plant's SSCs, and so forth. Specific issues of SMRs operation include how to control multi-module SMR plant from a single MCR and to cope with load following operation compared to conventional large nuclear power plant. Member States interested in SMRs need up-to-date information and technical know-how on the operational readiness and operational safety of SMRs prior to commercial operation.

Some features of existing large nuclear power plants such as a combined MCR for 4 (four) units of CANada Deuterium Uranium (CANDU) reactor of the Darlington nuclear power plant in Canada, and flexible operation with combination of Electricité de France (EDF) nuclear fleet in France and digitized HMI in the Advanced Power Reactor 1400 (APR1400) of the Saeul nuclear power plant Unit 1 and 2 in the Republic of Korea are similar to characteristics of early deployable SMRs. Therefore, comparing those features between large nuclear power plants and SMRs will be easy means to understand features of SMRs.

Collection and dissemination of existing operational design concepts and practical procedures on currently operating and near term deployable SMRs can preclude potential hazards arising during commissioning and operating SMRs. Over the past several years, the IAEA and other international organizations have provided publications regarding SMRs. However, operational readiness for SMRs is not yet reflected comprehensively within existing resources.

1.2. OBJECTIVE

The objective of this publication is to provide state-of-the art technical information on the approach and preparation for operation of near term deployable SMRs in comparison to related experience and knowledge gathered from commercial nuclear power plants using large water cooled reactors with many years of operating experience. This publication considers near term deployable SMRs as reactor models with electrical output of 300 MW(e) or less and planned to be connected to the electric grid between 2020 to 2035. As of 2025, only two nuclear power plants based on SMR are in commercial operation: a PWR based FNPP and a high temperature gas cooled reactor. There are, however, about twenty other SMR designs expected to start construction and connect to the electric grid by 2035, as first of a kind (FOAK) units [1]. This publication wi focus on eight (8) of the near term deployable SMR models, covering their designs and technologies in operational perspectives.

1.3. SCOPE

The scope of this publication is the identification of characteristics of near term deployable SMRs in terms of the approach and preparation for operation in comparison to that of conventional large nuclear power plants from the viewpoint of plant operators. The emphasis here is on the activities of operators of SMRs and large nuclear power plants who apply the technical specifications including modes of plant operation, OLCs after the first new fuel assembly is loaded into a reactor core; and perform the primary system heat-up and plant start-up operation including integrated commissioning tests through operating procedures and guidelines before their commercial operation. This report also provides how to manipulate plant control systems and plant power during baseload and/or flexible operation, how to assure fuel integrity, and how to optimize the arrangement of an MCR for a single and/or multi-module power plant including their HMI.

1.4. STRUCTURE

This publication comprises seven sections. This section 1 presents the background, objective, scope and structure of the publication. Section 2 provides key features in near term deployable SMRs, addresses modularity, flexible power generation and non-electric applications as a means to acknowledge how SMRs are utilized. Section 3 discusses key aspects of plant operations composed of modes of operation, OLCs in technical specification, start-up and operational approach, and operating procedures and guidelines to inform how SMRs are similar and on the other hand different to large nuclear power plants. Section 4 provides SMR control and manoeuvre approach addressing control of nuclear steam supply system (NSSS) that presents how the primary side systems and power control systems are being adequately controlled during operation. It also addresses operational approach in flexible operation of SMR and large nuclear power plants. Section 5 presents an overview of fuel integrity of SMRs explaining general fuel integrity in SMRs and provides fuel integrity during flexible operation in SMRs and addresses fuel characters of each SMR presented. Section 6 provides operating personnel management and HMI of SMRs in comparison to large nuclear power plants, addresses arrangement of MCR crew and roles and responsibilities of each individual in a MCR as well as HMI management. Section 7 concludes with a summary results and conclusion.

2. THE PERSPECTIVE OF NEAR TERM DEPLOYABLE SMALL MODULAR REACTORS

2.1. KEY FEATURES IN NEAR TERM DEPLOYABLE SMALL MOLDULAR REACTORS

Operational technical details of near term deployable SMRs in this section highlight their significant operational characteristics, parameters, and specified values as outlined in the publication. Additionally, the developmental progress and deployment milestones of SMRs will offer insights into their expected timelines. There are two distinct categories of SMRs such as water cooled SMRs consisting of eight different models, and other than water cooled SMRs, exemplified by the HTR-PM design.

2.1.1. Water cooled type small modular reactors

Water cooled type SMRs have certain advantages of being able to leverage proven technologies, extensive operational experience, and valuable lessons learned from the commercial fleets of large water cooled nuclear power plants. This includes insights gained from major safety events that have significantly influenced safety protocols and design considerations in the nuclear industry. However, despite the advantages, the design and deployment of water cooled SMRs are not without their specific challenges. Careful technical considerations need to be taken into account during their design, deployment and operation stages.

2.1.1.1. *ACP100 key design features*

The ACP100 is a SMR developed by China National Nuclear Corporation based on existing PWR technology. The plant layout is designed to be compact and flexible. The main technical parameters include a thermal power of 385 MW(t) and an electric power of 125 MW(e). Figure 1 shows the major design parameters of ACP100 [1]. The technical features include a passive safety system and high level of automation. The ACP100 continues its construction according to the following development and deployment milestones presented in the Table 1.

Parameter	Value
Reactor type	Integral PWR
Coolant/moderator	Light water / Light water
Thermal/electrical capacity, MW(t)/MW(e)	385 / 125
Primary circulation	Forced circulation
NSSS Operating Pressure (primary/secondary), MPa	15 / 4.6
Core Inlet/Outlet Coolant Temperature (°C)	286.5 / 319.5
Fuel type/assembly array	UO_2 / 17 × 17 square pitch
Number of fuel assemblies in the core	57
Fuel enrichment (%)	< 4.95
Core Discharge Burnup (GWd/ton)	< 52
Refuelling Cycle (months)	24
Reactivity control mechanism	Control rod drive mechanism (CRDM), Gd_2O_3 solid burnable poison and soluble boron acid
Approach to safety systems	Passive

FIG. 1. Major design parameters of ACP100(Source: CNNC, China, with permission).

TABLE 1. DEVELOPMENT AND DEPLOYMENT MILESTONES OF ACP100

Year/Date	Milestone/Description
2018	Achievement of preliminary safety assessment report (PSAR)
2020	Application for authorization at Changjiang nuclear power site, Hainan, China
2021.7	First concrete date (FCD) on Changjiang nuclear power site, Hainan, China

2025.4	Target start of cold test
2025.12	Target first fuel load
2026.3	Target first grid connection
2026.5	Target commercial operation

2.1.1.2.   BWRX-300 key design features

The BWRX-300 is designed by GE Hitachi Nuclear Energy. It employs natural circulation and passive safety, leveraging the design and licensing development of Economic Simplified Boiling Water Reactor (ESBWR), while introducing further simplification and safety enhancements. The BWRX-300 is a light water cooled reactor generating approximately 300 MW(e).

The passive safety features provide mitigation for transients and accidents with no reliance on operator actions, external cooling water, or AC generated power. It includes the capability to remove decay heat for a period in excess of seven days if there were a complete loss of all electrical power. This heat removal function can be extended indefinitely by simple actions to provide makeup water to the installed water pools. The BWRX-300 safety strategy includes a robust defense in depth approach that aligns with the principles of IAEA SSR-2/1, Safety of Nuclear Power Plants: Design. This layered approach to safety mitigates postulated event sequences in design basis and protects against common cause failures.

The structures and systems of the BWRX-300 are designed for efficient, cost effective construction, operation, and maintenance with a focus on safety. The reactor building is embedded, providing significant protection for natural and other hazards, while the other buildings are simple surface mounted structures.

The BWRX-300 is a direct cycle boiling water reactor and therefore provides for a simple operational concept while achieving a high degree of safety. The design leverages the operating experience of the fleet of existing BWRs as well as the extensive test and analytical programmes completed during the development of the Simplified Boiling Water Reactor (SBWR) and ESBWR plants. The natural circulation of the light water reactor coolant has been proven through years of successful operation of the Dodewaard nuclear plant as well as the performance of forced circulation BWRs during periods without pump operation. Figure 2 shows the major design parameters of BWRX-300 [1]. The BWRX-300 continues its development according to the following development and deployment milestones as listed in the Table 2.

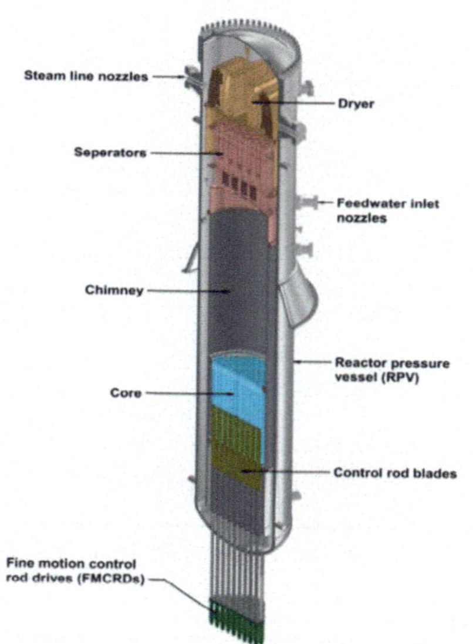

Parameter	Value
Reactor type	Boiling water reactor
Coolant/moderator	Light-water / Light-water
Thermal/electrical capacity, MW(t)/MW(e)	870 / 300
Primary circulation	Natural circulation
NSSS Operating Pressure (primary/secondary), MPa	7.2 / Direct cycle
Core Inlet/Outlet Coolant Temperature (°C)	270 / 288
Fuel type/assembly array	UO_2 / 10×10 array
Number of fuel assemblies in the core	240
Fuel enrichment (%)	3.81 (avg) / 4.95 (max)
Core Discharge Burnup (GWd/ton)	12–24
Refuelling Cycle (months)	49.6
Reactivity control mechanism	Rods and solid burnable absorber (B_4C, Hf, Gd_2O_3)
Approach to safety systems	Fully passive

FIG. 2. *Major design parameters of BWRX-300. (Source: GE Hitachi Nuclear Energy, United States, with permission).*

TABLE 2. DEVELOPMENT AND DEPLOYMENT MILESTONES OF BWRX-300

Year/Date	Milestone/Description
2017	Start of conceptual design
2019	Initiation of pre-application activities with the US NRC
2020	Initiation of phase 1 and 2 vendor design review (VDR) with Canada Nuclear Safety Commission in Canada
2021	Selection of the BWRX-300 by OPG after extensive evaluation of all SMRs
2022	Target OPG submittal of License to Construct for DNNP-1
2028	Target commercial operation of lead BWRX-300

2.1.1.3. CAREM key design features

The Central ARgentina de Elementos Modulares (CAREM) is an integral PWR type SMR developed by Comisión Nacional de Energía Atómica (CNEA), Argentina and its national consortium. Based on light water reactor (LWR) technology, CAREM adopts design simplification through the integration of NSSS for the reactor to achieve a higher level of safety. The power level is optimized for areas and users with small electricity grids. Essential design features of CAREM include integrated primary cooling system; self-pressurization; natural circulation for core cooling; in-vessel control rod drive mechanisms; and passive safety systems.

One unit of CAREM generates 100 MW(t) and up to 30 MW(e) per unit. Being an integral PWR type SMR, its primary circuit and components, i.e. steam generators are installed inside the reactor vessel. The primary natural circulation, and it does not need any primary recirculation pumps. The self-pressurization is achieved by balancing steam production and condensation in the vessel, without a separate pressurizer vessel. Figure 3 shows the major design parameters of CAREM25 [1]. The CAREM25 continues its construction according to the following development and deployment milestones as listed in Table 3.

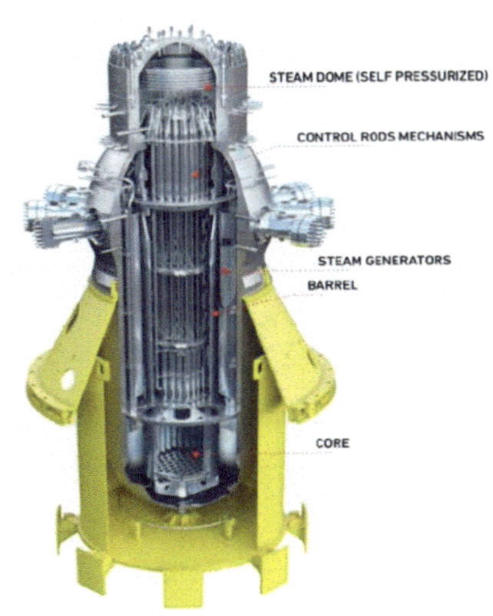

Parameter	Value
Reactor type	Integral PWR
Coolant/moderator	Light water / Light water
Thermal/electrical capacity, MW(t)/MW(e)	100 / ~30
Primary circulation	Natural circulation
NSSS Operating Pressure (primary/secondary), MPa	12.25 / 4.7
Core Inlet/Outlet Coolant Temperature (°C)	284 / 326
Fuel type/assembly array	UO_2 pellet/hexagonal
Number of fuel assemblies in the core	61
Fuel enrichment (%)	3.1%
Core Discharge Burnup (GWd/ton)	24
Refuelling Cycle (months)	14
Reactivity control mechanism	Control rod driving mechanism (CRDM) only
Approach to safety systems	Passive

FIG. 3. Major design parameters of CAREM25. (Source: CNEA, Argentina, with permission).

TABLE 3. DEVELOPMENT AND DEPLOYMENT MILESTONES OF CAREM25

Year/Date	Milestone/Description
1984	Launch of the official CAREM project by CNEA
2006	Listing CAREM25 project on Argentina Nuclear Reactivation Plan among priorities of national nuclear development
2009	Submission of PSAR for CAREM25
2011	Beginning of site excavation

2013	Approval of construction license
2014	Start of civil works
2029	Planned first criticality

2.1.1.4. KLT-40S key design features

The KLT-40S is a type of marine based reactor like those used on icebreakers and developed by JSC 'Afrikantov OKBM' in Russian Federation to produce thermal power of 150 MW(t) and electrical power of 35 MW(e) per module. The Akademik Lomonosov is a FNPP consisting of a reactor vessel and a floating power unit equipped with two (2) KLT-40S reactor systems. It provides a district heating of 58.15 MW(t) and electric power of about 70 MW(e) in the aggregate. The FNPP building is fully welded, divided by partitions into compartments forming residential and technological blocks.

Figure 4 shows the major design parameters of the KLT-40S [1]. After the regulator permitted operation of the onshore electric substation, comprehensive trial tests of the substation had been carried out. The FNPP has been commissioned for commercial operation, with thermal power supply to the Pevek district heating system. The Akademik Lomonosov achieved its commercial operation according to the following development and deployment milestones as listed in Table 4.

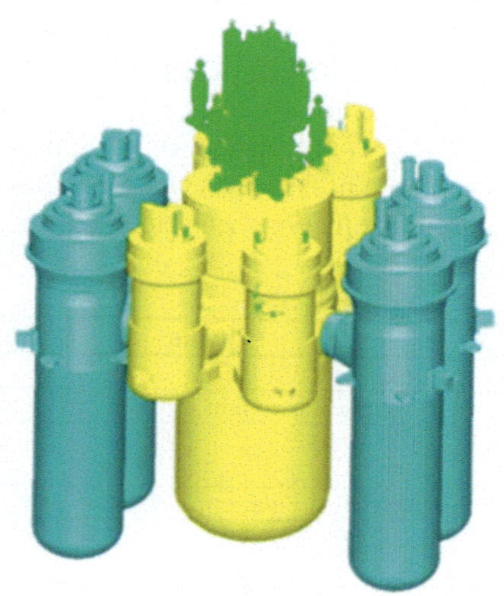

Parameter	Value
Reactor type	PWR
Coolant/moderator	Light water / Light water
Thermal/electrical capacity, MW(t)/MW(e)	150 / 35
Primary circulation	Forced circulation
NSSS Operating Pressure (primary/secondary), MPa	12.7 / 3.82
Core Inlet/Outlet Coolant Temperature (°C)	280 / 316
Fuel type/assembly array	UO_2 pellet in silumin matrix
Number of fuel assemblies in the core	121
Fuel enrichment (%)	18.6
Core Discharge Burnup (GWd/ton)	45.4
Refuelling Cycle (months)	30-36
Reactivity control mechanism	Control rod driving mechanism
Approach to safety systems	Active (partially passive)

FIG. 4. *Major design parameters of KLT-40S. (Source: Afrikantov OKBM, Russian Federation, with permission).*

TABLE 4. DEVELOPMENT AND DEPLOYMENT MILESTONES OF KLT-40S

	Milestone/Description
1998	Launch of the first FNPP project
2002	Approval of the environmental impact assessment
2017	Completion of construction and systems testing of the floating power unit
2018	Attainment of reactor's first criticality
2019.12	Connection to the grid in Pevek
2020.5	Commercial operation

2.1.1.5. Rolls-Royce SMR key design features

The Rolls-Royce SMR is being developed by the Rolls-Royce company in the United Kingdom generating a total thermal power of 1358 MW(t) with electrical power of 470 MW(e). It is a three loop PWR and is boron free for normal operation. Emergency boron injection is used as the second, independent means of reactivity control. Control rods provide the primary means of shutdown and hold down. It uses potassium hydroxide chemistry rather than lithium hydroxide based. The innovation is in how to design for modularization including standardization and road transport to deliver build certainty. The modular built within factories allows for certain testing and commissioning activities to take place off-site, within a controlled environment, reducing the on-site work. Systems or subsystems can be tested to accelerate and de-risk the on-site activities. Anticipating a conventional set of activities for hot operations and power ascension, with a desire to accelerate these if possible.

It can provide both base load and load following capability. The power station is designed to integrate with alternative applications, such as hydrogen production. Passive safety features using natural circulation provide the required levels of protection against intact circuit faults and loss of coolant accidents. Figure 5 shows the major design parameters of the Rolls-Royce SMR [1]. The Rolls-Royce SMR continues its development according to the following development and deployment milestones as listed in Table 5.

Parameter	Value
Reactor type	3-loop PWR
Coolant/moderator	Light water / Light water
Thermal/electrical capacity, MW(t)/MW(e)	1358 / 470
Primary circulation	Forced circulation
NSSS Operating Pressure (primary/secondary), MPa	15.5 / 7.8
Core Inlet/Outlet Coolant Temperature (°C)	295 / 325
Fuel type/assembly array	UO_2 / 17 × 17 Square
Number of fuel assemblies in the core	121
Fuel enrichment (%)	<4.95
Core Discharge Burnup (GWd/ton)	50–60
Refuelling Cycle (months)	18
Reactivity control mechanism	Control rods
Approach to safety systems	Passive and Active

FIG. 5. *Major design parameters of Rolls-Royce SMR. (Source: Rolls Royce Plc., United Kingdom, with permission).*

TABLE 5. DEVELOPMENT AND DEPLOYMENT MILESTONES OF ROLLS-ROYCE SMR

Year/Date	Milestone/Description
2015	Rolls-Royce development of initial reference design
2016	Formation of consortium for design of whole power station
2017	Mature design concept developed
2022	Formal regulation entered in the UK
2026	Projected start of first of a kind construction
2030	Planned first of a kind commercial operation

2.1.1.6. *SMART key design features*

The SMART is an integral pressurized water reactor developed by Korea Atomic Energy Research Institute (KAERI) in Republic of Korea. A standard design certification application for the SMART was approved by the Nuclear Safety and Security Commission (NSSC) in the Republic of Korea in 2012. To enhance safety and reliability, joint development of SMART was subsequently conducted by KAERI and King Abdullah City for Atomic and Renewable Energy (K.A.CARE). Then, KAERI, Korea Hydro & Nuclear Power Co., Ltd. and K.A.CARE submitted the standard design certification application for SMART 100 to the NSSC in 2020.

A unit of the SMART100 is designed to produce a rated electrical power of 107 MW(e) from a core power of 365 MW(t) and the nuclear power plant of a SMART100 accommodating two units can generate a rated electrical power of 214 MW(e). Each unit installed in its individual reactor containment and auxiliary building is connected to its individual turbine and generator. The two units share a common compound building. The SMART100 design aims at improvement in economics through system simplification, unit modularization, reduction of construction time and high plant availability. Hence, the SMART100 adopts advanced design features and passive safety systems with proven technologies. Figure 6 shows the major operational design parameters of the SMART100 [1]. The SMART100 continues its development according to the following development and deployment milestones presented in Table 6.

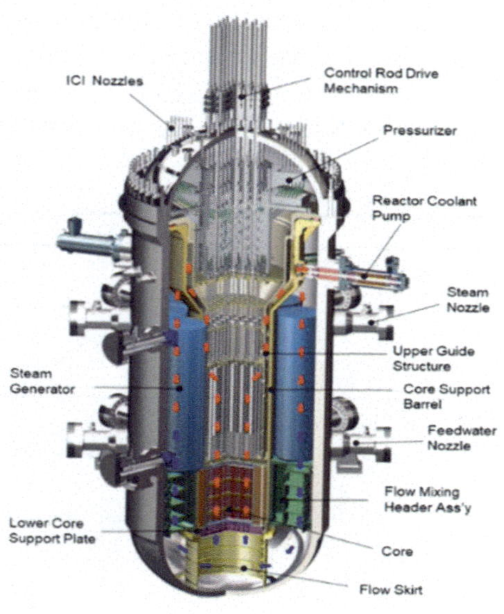

Parameter	Value
Reactor type	Integral PWR
Coolant/moderator	Light water / Light water
Thermal/electrical capacity, MW(t)/MW(e)	365 / 107
Primary circulation	Forced circulation
NSSS operating pressure (primary/secondary), MPa	15 / 5.8
Core inlet/Outlet coolant temperature (°C)	296 / 322
Fuel type/assembly array	UO_2 pellet / 17x17 square
Number of fuel assemblies in the core	57
Fuel enrichment (%)	< 5
Core discharge burnup (GWd/ton)	< 54
Refuelling cycle (months)	30
Reactivity control mechanism	Control rod drive mechanisms and soluble boron
Approach to safety systems	Passive

FIG. 6. Major design parameters of SMART. (Source: KAERI, Republic of Korea, with permission).

TABLE 6. DEVELOPMENT AND DEPLOYMENT MILESTONES OF SMART

Year/Date	Milestone/Description
2012.7	Completion of technology verification, SDA
2012.3	Launch of first step of post-Fukushima corrections and commercialization
2015.11	Start of pre-project engineering
2019.2	Completion of pre-project engineering
2020.1	SDA application for SMART100

2.1.1.7. *VOYGR^{TM} key design features*

VOYGR™ is an integral pressurized water reactor designed by NuScale Power LLC in the United States of America. A NuScale Power Module™ (NPM) includes the containment, pressurizer, steam generators, RPV, and reactor core in an integral package. The reactor core consists of an array of reduced-height LWR fuel assemblies and control rod clusters at standard enrichments. The simple design eliminatNPMes reactor coolant pumps (RCPs), large bore piping, and other systems and components found in large conventional reactors. The combined NPM and associated auxiliary systems such as the turbine and generator are referred to together as a unit. Multiple units configured together comprising the entire plant are referred to as VOYGR™. Various configurations such as 12-unit, 6-unit, or 4-unit would be referred to as VOYGR™-12, VOYGR™-6, or VOYGR™-4 respectively. Each NPM has key design features approved by the NRC including no operator action or AC/DC power needed to shut down reactors and no need to add water to keep reactors safe and cooled for an unlimited time. The VOYGR™ plant does not require connection to the grid or '1E power' for safety. This permits 'off-grid' operation: a very important feature for providing reliable power and process heat to industrial applications. The site can be located at the 'end of line' for a grid connection, distributed generation applications, coal plant repowering, and for district heating.

Figure 7 shows the major design parameters of the VOYGR™ [1]. The NuScale's simplified and standardized design will help to lower the cost of materials and manufacturing and support modular build gains. The VOYGR™ plant can support either wet cooled or air-cooled condenser systems that may be an option for customers where water scarcity is a concern. The VOYGR™ continues its development according to the following development and deployment milestones presented in Table 7.

NuScale submitted a design certification application for a VOYGR™-12 (50MW(e) per unit) design in 2016 approved by the US NRC in 2020. NuScale subsequently submitted a standard plant design certification application for a VOYGR™-6 (77MW(e) per unit) design in 2022. The standard plant design application is a stand-alone submittal following the NRC 10 CFR 52 'Licenses, Certifications, and Approvals for nuclear power plants' certification process. The standard plant design utilizes the design certification as the referenced basis and updated the content based on the power uprate and 6-unit design. The standard plant design focuses on safety and risk significance. Once the standard plant design is approved, it can be referenced by an applicant without repeating the review process.

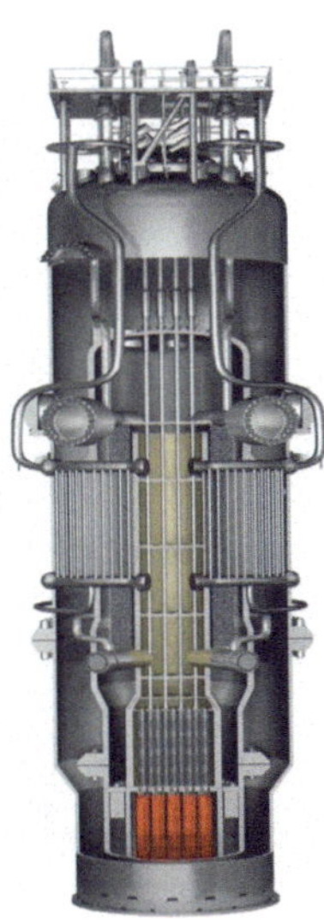

Parameter	Value
Reactor type	Integral PWR
Coolant/moderator	Light water / Light water
Thermal/electrical capacity, MW(t)/MW(e)	250 / 77 (gross)
Primary circulation	Natural circulation
NSSS Operating Pressure (primary/secondary), MPa	13.8 / 4.3
Core Inlet/Outlet Coolant Temperature (°C)	249 / 316
Fuel type/assembly array	UO_2 pellet / 17x17 square
Number of fuel assemblies in the core	37
Fuel enrichment (%)	≤ 4.95
Core Discharge Burnup (GWd/ton)	≥ 45
Refuelling Cycle (months)	Nominal 18
Reactivity control mechanism	Control rod drive, boron
Approach to safety systems	Control rod drive, boron

FIG. 7. *Major design parameters of VOYGR™. (Source: NuScale Power Inc., United States, with permission).*

TABLE 7. DEVELOPMENT AND DEPLOYMENT MILESTONES OF SMART

Year/Date	Milestone/Description
2008	Initiation of NRC Pre-Application
2013	U.S. DOE SMR Cooperative Agreement signed
2016	Submission of design certification application to the U.S. NRC
2020.9	Reception of SDA from the NRC
2022	Start of NPM production with the production of forging dies for the Upper RPV
2023	Submission of combined license application (COLA) to the NRC for first plant
2025	Construction targeted to start for first VOYGR-6 plant in Idaho, USA (Cancelled)

2.1.2. Other than water cooled type small modular reactors

Other than water cooled SMRs predominantly belong to the Generation IV category of nuclear reactors. These reactors are characterized by their advanced engineered features and an enhanced safety performance largely attributed to their inherent and passive safety features. There is a wide variety of other than water cooled SMRs, each with their own unique set of advantages and disadvantages. This includes high temperature gas cooled reactors, fast reactors, and molten salt reactors among others. Given the unique characteristics of each type of other than water cooled SMR, they may each require different safety standards and regulations. The widespread adoption of other than water cooled SMRs depends on their ability to compete economically with other energy sources. This requires not only technological advancements but also strategic planning and supportive policies.

2.1.2.1.   HTR-PM key design features

The HTR-PM owned by Huaneng Group Co., Ltd. in China is a modular demonstration plant composed of two standard NSSS modules connected to a conventional steam turbine generator developed by Institute of Nuclear and New Energy Technology of Tsinghua University. The plant layout is designed to be modular and flexible. The main technical parameters include a thermal power of 250 MW(t) and an electric power of 210 MW(e). The technical features include a high level of safety and a high level of automation. Figure 8 shows the major design parameters of the HTR-PM [1].

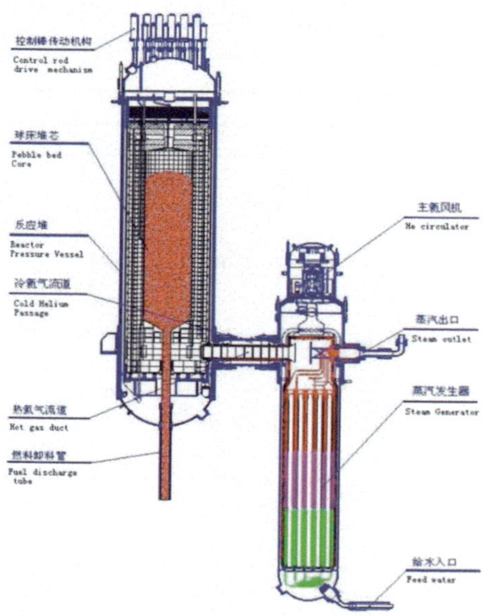

Parameter	Value
Reactor type	Modular pebble bed high temperature gas cooled reactor
Coolant/moderator	Helium / Graphite
Thermal/electrical capacity, MW(t)/MW(e)	2 × 250 / 210
Primary circulation	Forced circulation
NSSS Operating Pressure (primary/secondary), MPa	7 / 13.25
Core Inlet/Outlet Coolant Temperature (°C)	250 / 750
Fuel type/assembly array	Spherical elements with coated particle fuel
Number of fuel assemblies in the core	420 000 (in each reactor module)
Fuel enrichment (%)	8.5
Core Discharge Burnup (GWd/ton)	90
Refuelling Cycle (months)	On-line refuelling
Reactivity control mechanism	Control rod insertion
Approach to safety systems	Combined active and passive

FIG. 8. Major design parameters of HTR-PM. (Source: Huaneng Group, China, with permission).

The HTR-PM accomplished its commercial operation according to the following development and deployment milestones as listed in Table 8.

TABLE 8. DEVELOPMENT AND DEPLOYMENT MILESTONES OF SMART

Year/Date	Milestone/Description
2006–2008	Completion of Basic design of HTR-PM
2012.12	First concrete date (FCD) of HTR-PM
Q4/2020	Startup commissioning test of primary circuit
2021.8	First reactor core load
2021.9	First criticality achievement of first reactor
	Second reactor core load
2021.12	First reactor grid connection
2022.9	Second reactor grid connection
2022.10	Highest power level achievement of first reactor
2022.12	Highest power level achievement of second reactor
2023.12	Start of commercial operation of HTR-PM

The HTR-PM 600 high temperature gas cooled reactor nuclear power plant is a follow-up commercial version of the HTR-PM demonstration plant. Six (6) HTR-PM reactor modules, in parallel, are designed to connect to one steam turbine to form a 600 MW(e) power plant having a control room with fewer HMI devices and fewer operators.

2.2. MODULARITY

The modularity of SMRs is a key feature that contributes to their potential for enhanced safety, security, and economic efficiency. The modularity in SMRs refers to their scalability and factory fabrication of major components of the integral NSSS. This can undertake assembled and/or tested structures, systems to be transported to the site that can support reducing on-site preparation and decreasing construction time. Many SMR designs rely on a modular system, allowing customers to simply add modules to achieve a desired output. This modularity also provides benefits such as lower initial capital investment, greater scalability, and siting flexibility for locations unable to accommodate more traditional larger reactors.

The scalability of SMRs is to have the ability to adjust power output to meet changing demands accomplished through modular designs allowing the addition or removal of modules to accommodate energy demands. SMRs can be deployed in diverse locations, including remote or off-grid areas but also enable a more diverse energy mix. SMRs can also be built in smaller increments, decreasing the initial capital investment required. This reduces the financial risk associated with large scale nuclear projects. Due to their smaller size and off-site manufacturing, SMRs can be built and operational more quickly than traditional nuclear power plants. This faster deployment time can support countries meeting their energy demand in a timely manner and assist the transition in a low carbon economy.

For VOYGR™ plants fabrication of the NSSS, primary plant systems, and many balance of plant systems will be performed in a factory setting. A smaller plant footprint, fewer components per plant, and a high percentage of modular factory fabricated components and packages all help to reduce construction complexity and duration. Quality assurance, inspection, and testing activities performed at the factory will also help to reduce potential impacts on the construction schedule and improve cost certainty.

The multi-module nature of a plant that can contain up to 12 separate modules and power conversion systems operating independently allows:

— The VOYGR™ plant will continue to provide power on a continuous basis even when individual modules are taken offline for refuelling or maintenance;
— Modules can also be returned to service one at a time to match the demand of the offsite grid in 77 MW(e) increments to help black start the grid when power is ready to be restored;
— The VOYGR™ plant has been designed so one or more modules can provide house load in the case of a loss of offsite power.

Due to this redundancy in design, power output from a VOYGR™ power plant can be assured albeit at a reduced total power level, throughout the lifetime of the plant.

2.3. FLEXIBLE OPERATION OF NUCLEAR POWER PLANTS

Flexible operation in nuclear power plants, in other words, non-baseload operation, refers to the ability of these nuclear units to adapt their electrical output to match the electrical demand and control the frequency of the electrical system. The term 'flexible operation' is used to describe any mode of operation that is not baseload, specifically focusing on changing electrical output to match generation with demand on the electrical grid system [2]. Demand on flexible operation in nuclear power plants has been increasing especially due to growth in renewable energy, and changes in the electric supply system and market so newly built nuclear power plants including SMRs are generally designed with flexible operation capable of conducting load following and frequency control that are two important concepts in power system operation and management.

The load following is to vary the output of a generating unit in a planned way, or in response to an instruction or control signal from the grid control centre, reducing the output when the load (electrical demand) on the system is reduced (e.g., at night, at weekends and on public holidays) and increasing the output to maximum when the electrical demand is high. Automatic frequency control is a method of operating a generating unit at less than full output, under automatic control, so that its output increases automatically if the system frequency falls and decreases automatically if the system frequency rises [2]. Any impacts on the reactor core integrity need to be secured with utmost care, according to all safety regulations and guidelines.

2.4. NON-ELECTRIC APPLICATIONS

Non-electric applications powered by SMRs could present sustainable solutions for several energy challenges current and future generations will have to face. The following is a list of non-electric applications that SMR technology could be leveraged:

— Hydrogen production;
— Desalination;
— Power carbon capture & sequestration facilities with 100% clean power;
— Direct air capture;
— Decarbonizing industrial heat for steelmaking and manufacturing; Steam electrolysis; Methane reforming; Petroleum refining; Ethylene and styrene production; District heating; Pulp and paper production; Oil production from shale and tar sands; Glass and cement manufacture, etc.

The BWRX-300 can be adapted for non-electric applications in addition to the electrical generation function. Possible non-electric applications include the use of the generated steam to serve the district heating function, hydrogen production, or direct air capture of CO_2. Studies were carried out in Argentina and other countries to evaluate the capability and competitiveness of CAREM25 for non-electrical applications such as seawater desalination and hydrogen production with positive results. The 'Akademik Lomonosov' FNPP with KLT-40S reactor generates heat in the amount of up to 50 Gcal/h (58.11 MW(t)) and transfers it to the shore, where it is used to heat buildings and structures. Up to now it covers 100% of the heat demand in Pevek. It can also be used for desalination purpose to provide 20 000–100 000 m^3/h of potable water[1].

Through a feasibility study between KAERI in Republic of Korea and K.A.CARE in Saudi Arabia to couple the SMART100 with a desalination plant, a unit of the SMART100 was confirmed to have a capacity to generate electricity of 100 MW(e) for a population of 100 000 along with fresh water of 40 000 tonnes per day. In that study, the desalination system consists of four multi-effect distillation (MED) units, each with a thermal vapour compressor (MED-TVC). The SMART100 and MED-TVC units are connected through the steam transformer. The steam transformer produces the motive steam by using the extracted steam from a turbine and supplies the process steam to the desalination plant.

Availability of the SMART100 for oil sand recovery is being studied. The thermal oil sand recovery processes, such as the steam assisted drainage gravity (SAGD) and cyclic steam stimulation (CSS), require a lot of high pressure and temperature steam to enable the oil sand to be soaked. A conceptual design of the SMART100 for oil sand recovery is equipped with the mid-loop system between the SMART100 and the previous oil sand process to form an additional radioactive material barrier against the environment and separate their water chemistry processes.

VOYGR™ plants can help companies meet sustainability goals by providing carbon free heat for manufacturing processes. One module of VOYGR can produce 250 MW(t) of heat for industrial applications such as chemical processing, enhanced oil recovery and production of synthetic fuels, including hydrogen. Furthermore, this heat can be utilized in a range of low temperature, low pressure applications such as water desalination and district heating. For instance, a single module of VOYGR coupled to a desalination plant can produce 291 ML/d of potable water. From 250 MWt 2053 kg/h of hydrogen or nearly 50 metric tons per day can be produced.

---

[1] When used in a desalination complex (floating power unit + desalination unit)

3. KEY ASPECTS OF NUCLEAR POWER PLANT OPERATIONS

3.1. MODES OF NUCLEAR POWER PLANT OPERATION

The mode of operation in every nuclear power plant including SMRs refers to the plant's operating status classified under one of several 'modes'. These modes are important as they determine how a reactor responds during all modes of operation. The complete details of these modes can be found in the standard technical specification.

(a) APR1400 modes of operation

Saeul nuclear power plant unit 1 in commercial operation since December 2016 is one of the APR1400 models as a LWR type located in Saeul, Republic of Korea producing a thermal power of 4000 MW(t) and an electric power of 1455 MW(e) that has been submitted for standard design certification application by Korea Electric Power Corporation and Korea Hydro & Nuclear Power Co., LTD.

Saeul nuclear power plant unit1, as an example for conventional large nuclear power plants, has 6 modes of operation defined in the technical specification. Each mode has to correspond to any one inclusive combination of core reactivity condition, power level, reactor coolant cold leg temperature, and reactor vessel head closure bolt tensioning specified in the below Table 9 with fuel in reactor vessel.

TABLE 9. MODES OF OPERATION FOR APR1400

Mode	Title	Reactivity condition (Keff)	Rated thermal power (%)[a]	Reactor cold leg coolant temperature (T_{cold}/°C)
1	Power operation	≥ 0.99	> 5	n.a.[d]
2	Startup	≥ 0.99	≤ 5	n.a.[d]
3	Hot standby	< 0.99	n.a.[d]	≥ 177
4	Hot shutdown[b]	< 0.99	n.a.[d]	$99 < T_{cold} < 177$
5	Cold shutdown[b]	< 0.99	n.a.[d]	≤ 99
6	Refuelling[c]	n.a.[d]	n.a.[d]	n.a.[d]

[a] Excluding decay heat
[b] All reactor vessel head closure bolts fully tensioned
[c] One or more reactor vessel head closure bolts less than fully tensioned
[d] Not Applicable

Each mode of plant operation is defined as follows:

— Mode 1 'Power Operation' defined as core critical and rated thermal power more than 5% for synchronization of turbine/generator to the NSSS;
— Mode 2 'Startup' defined as core critical and rated thermal power below or equal to 5%;
— Mode 3 'Hot Standby' defined as core subcritical and T_{cold} higher than or equal to 177 °C;
— Mode 4 'Hot Shutdown' defined as core subcritical and T_{cold} higher than 99°C and below 177 °C where safety systems can bring the reactor;
— Mode 5 'Cold Shutdown' defined as core subcritical and T_{cold} below or equal to 99 °C where the reactor can reach with additional shutdown cooling function;
— Mode 6 'Refuelling' defined where one or more stud bolts and reactor closure head are not fully tensioned for refuelling and maintenance.

(b) BWRX-300 modes of operation

The BWRX-300 has five modes of operation as in the below Table 10.

TABLE 10. MODES OF OPERATION FOR BWRX-300

Mode	Title	Reactivity status	Average reactor coolant temperature (°C)
1	Power operation	≥ 10% rated thermal power	n.a.[d]
2	Startup	< 10% rated thermal power control rods capable of withdrawal	n.a.[d]
3	Hot shutdown	Not more than one control rod pair capable of withdrawal	> 93.3
4	Cold shutdown[a]	Not more than one control rod pair capable of withdrawal	≤ 93.3
5	Refuelling[b,c]	n.a.[d]	n.a.[d]

[a] All Reactor Pressure Vessel (RPV) head closure studs fully tensioned
[b] One or more RPV head closure studs less than fully tensioned
[c] Reactivity state conditions for fuel movement and test configurations located within applicable Technical Specifications
[d] Not Applicable

The technical specifications of BWRX-300 define six five (5) modes of operation in aligned with passive plant guidance. These operational modes are defined as follows:

— Mode 1 'Power Operation' is defined with the Reactor Mode Switch in the 'RUN' position at any average reactor coolant temperature with reactivity ≥ 10% rated thermal power;
— Mode 2 'Startup' is defined with the Reactor Mode Switch in the 'STARTUP' position at any average reactor coolant temperature. Startup Operation is also defined with the Reactor Mode with reactivity < 10% rated thermal power control rods capable of withdrawal;
— Mode 3 'Hot Shutdown' is defined with the Reactor Mode Switch in the 'SHUTDOWN' position at an average reactor coolant > 93.3°C and all RPV head closure studs fully tensioned; Not more than one control rod pair capable of withdrawal;
— Mode 4 'Cold Shutdown' is defined with the Reactor Mode Switch in the 'SHUTDOWN' position at an average reactor coolant temperature ≤ 93.3°C. In this mode, the temperature of the reactor coolant system (RCS) is low, and the isolation condensers are not functional. Decay heat removal is accomplished through the shutdown cooling system (SDC);
— Mode 5 'Refuelling' is defined where one or more RPV head closure studs less than fully tensioned. Reactivity state conditions for fuel movement and test configurations located within applicable Technical Specifications.

(c) KLT-40S modes of operation

For the Akademik Lomonosov FNPP, the KLT-40S reactor has five modes of operation as in the below Table 11.

TABLE 11. MODES OF OPERATION FOR KLT-40S

Mode	Title	Reactivity condition (Keff)	Reactor cold/hot leg coolant temperature (°C)
1	Power operation	≥ 0.99	280/316
2	First criticality	≥ 0.99	280/316
3	Hot shutdown	< 0.99	> 99

| 4 | Cold shutdown | < 0.99 | ≤ 99 |
| 5 | Shutdown for repair/refuelling | n.a.[a] | ≤ 60 |

[a] Not Applicable

Each mode of plant operation of the KLT40s is defined as follows:

— Mode 1 'Power operation' defined as reactor operation at power starting from auxiliary (house load) one up to the rated one;
— Mode 2 'First criticality' defined as reactor works at the minimum controllable power level when all the parameters of the primary circuit are equal to the power operation parameters;
— Mode 3 'Hot shutdown' defined as the reactor is shut down, but pressure is equal to the operation one of 12.7 MPa and temperature varies from 100 °C to 280 °C;
— Mode 4 'Cold shutdown' defined reactor is shut down; the temperature is less than or equal to 99 °C, and the pressure can vary;
— Mode 5 'Shutdown for repair/refuelling' defined as the reactor is shut down; the temperature is less than or equal to 60 °C and the pressure equal to atmospheric.

(d) SMART100 modes of operation

For the SMART100, six modes of operation are defined in the technical specification. Each mode corresponds to any one inclusive combination of core reactivity condition, power level, reactor coolant average temperature and reactor closure head and stud bolt tensioning as follows in below Table 12.

TABLE 12. MODES OF OPERATION FOR SMART100

Mode	Title	Reactivity condition	Rated thermal power (%)[a]	Reactor coolant average temperature (T_{avg}/ °C)
1	Power operation	≥ 0.99	> 20	n.a.[d]
2	Startup operation	≥ 0.99	≤ 20	n.a.[d]
3	Hot standby	< 0.99	n.a.[d]	> 215
4	Safe shutdown	< 0.99	n.a.[d]	$93 < T_{avg} ≤ 215$
5	Cold shutdown[b]	< 0.99	n.a.[d]	≤ 93
6	Refuelling operation[c]	n.a.[d]	n.a.[d]	n.a.[d]

[a] Excluding decay heat
[b] All stud bolts and reactor closure head are fully tensioned
[c] One or more stud bolts and reactor closure head are not fully tensioned
[d] Not Applicable

Each mode of plant operation is defined as follows:

— Mode 1 'Power Operation' defined as core critical and rated thermal power more than 20% for synchronization of turbine/generator to the NSSS;
— Mode 2 'Startup Operation' defined as core critical and rated thermal power below 20%;
— Mode 3 'Hot Standby' defined as core subcritical and reactor coolant average temperature higher than 215 °C;
— Mode 4 'Safe Shutdown' defined as core subcritical and reactor coolant average temperature higher than 93°C and below 215 °C where safety systems can bring the reactor;

— Mode 5 'Cold Shutdown' defined as core subcritical and reactor coolant average temperature below 93 °C where the reactor can reach with additional shutdown cooling function;
— Mode 6 'Refuelling Operation' defined where one or more stud bolts and reactor closure head are not fully tensioned for refuelling and maintenance.

(e) VOYGR^{TM} modes of operation

The VOYGR™ has set out five modes of operation. Each mode corresponds to any one inclusive combination of core reactivity condition, reactor coolant temperature, control rod assembly, chemical volume control system (CVCS), containment flooding and drain system, and reactor vent valve and reactor vessel flange bolt tensioning specified in the below Table 13. With fuel in reactor vessel.

TABLE 13. MODES OF OPERATION FOR VOYGRTM

Mode	Title	Reactivity condition (k_{eff})	Other condition
1	Operations	≥ 0.99	N/A
2	Hot shutdown	< 0.99	Any indicated reactor coolant temperature ≥ 174 °C (345 °F) <u>AND</u> Not passively cooled
3	Safe shutdown[a]	< 0.99	All indicated reactor coolant temperature < 174 °C (345 °F) <u>OR</u> Not passively cooled
4	Transition[b, c]	< 0.95	N/A
5	Refuelling[d]	N/A	N/A

[a] Any control rod assembly capable of withdrawal, or any CVCS or containment flooding and drain system connection to the module not isolated.
[b] All control rod assemblies incapable of withdrawal, and CVCS and containment flooding and drain system connections to the module isolated, and all reactor vent valves electrically isolated.
[c] All reactor vessel flange bolts fully tensioned.
[d] One or more reactor vessel flange bolts less than fully tensioned

Each mode of plant operation is defined as follows:

— Mode 1 'Operation' defined as the reactor critical;
— Mode 2 'Hot Shutdown' defined as the reactor shutdown and reactor coolant temperature greater than or equal to 174 °C (345 °F) and not passively cooled;
— Mode 3 'Safe Shutdown' defined as the reactor shutdown, and reactor coolant temperature less than 174 °C (345 °F) OR Passively cooled and any control rod assembly capable of withdrawal, or CVCS or containment flooding and drain system connection to Module not isolated;
— Mode 4 'Transition' defined where the NPM is isolated from support systems with control rods incapable of being moved and all reactor vessel flange bolts fully tensioned;
— Mode 5 'Refuelling' defined where the reactor is shut down and one or more reactor vessel flange bolts are less than fully tensioned.

(f) HTR-PM modes of operation

The HTR-PM has established five operational modes. Each mode corresponds to any one inclusive combination of core reactivity condition, power level, reactor coolant average temperature specified in the below Table 14.

TABLE 14. MODES OF OPERATION FOR HTR-PM

Mode	Title	Reactivity (k_{eff})	Rated thermal power (%)	Average core temperature ($T_{avg}/°C$)
1	Power operation	≥ 0.99	≥ 30	N/A
2	Startup	≥ 0.99	< 30	N/A
3	Normal shutdown	< 0.99	N/A	≥ 150
4	Cold shutdown	< 0.99	N/A	$150 > T_{avg} \geq 50$
5	Maintenance Shutdown	< 0.99	N/A	< 50

Each mode of plant operation is defined as follows:

— Mode 1 'Power Operation' defined as core critical and rated thermal power higher than or equal to 30%;
— Mode 2 'Startup' defined as core critical and rated thermal power below 30%;
— Mode 3 'Normal Shutdown' defined as core subcritical and average temperature of core higher than or equal to 150 °C;
— Mode 4 'Cold Shutdown' defined as core subcritical and average temperature of core less than 150 °C, higher than or equal to 50 °C;
— Mode 5 'Maintenance Shutdown' defined as core subcritical and average temperature of core below 50 °C.

3.2. OPERATIONAL LIMITS AND CONDITIONS IN TECHNICAL SPECIFICATION OF NUCLEAR POWER PLANTS

OLCs are used as limiting conditions for operation in the other term are a set of parameters with limiting values that are necessary for the safe operation of the plant [3] [4]. The OLCs are usually defined in the plant's technical specifications establishing requirements for items such as safety limits, limiting safety system settings, limiting control settings, limiting conditions for operation, surveillance requirements, design features, and administrative controls.

The OLCs are crucial for maintaining the safety and integrity of nuclear power plants during operation. They ensure that the plant operates within safe parameters so that any deviations are promptly addressed to prevent any uncontrolled release of radioactivity.

3.2.1. Safety limits

The concept of safety limits is based on the prevention of unacceptable releases of radioactive substances from the plant through the application of limits imposed on the temperatures of fuel and fuel cladding, and on the coolant pressure, pressure boundary integrity and other operational characteristics influencing the release of radioactive substances from the fuel [5]. Safety limits in nuclear power plants are defined as limits on operational parameters that an authorized facility has been shown to be safe within. These limits are not only established to prevent potential hazards on the reactor core but also factored into the design of nuclear power plants. If these safety limits are exceeded, the reactor has to be shut down. This ensures that nuclear power remains a safe means of generating electricity, with the risk of accidents being low and declining.

(a) APR1400 operational safety limits

In an APR1400, safety limits are categorized into two main areas: (i) reactor core safety limit and (ii) RCS pressure safety limit.

The reactor core safety limit applies to the core of the reactor, where the nuclear reactions occur. In modes 1 and 2 of operation, the following conditions has to be met:

— The departure from nucleate boiling ratio, a measure of how close the reactor is to the onset of boiling, has to be maintained at or above 1.29. This ensures that the reactor operates in a safe regime where the coolant does not transition into a steam phase that could potentially damage the fuel rods;
— The peak fuel centreline temperature has to be kept below 2,805°C. This limit decreases by 7.6°C per 10,000 MWD/MTU and is adjusted for burnable poison. This ensures that the fuel rods do not overheat, that could lead to fuel rod failure.

RCS pressure safety limit pertains to the pressure within the RCS crucial for maintaining the integrity of the system and preventing leaks. In modes 1, 2, 3, 4, and 5, the RCS pressure has to be maintained at or below 193.3 kg/cm^{3}A (19 MPa).

These safety limits are critical for the safe operation of the reactor and need to be strictly adhered to.

If the reactor unit violates either of the two (2) safety limits in the standard technical specifications, the following procedures has to be enacted:

— In the event of a reactor core safety limit violation, the unit is required to transition to mode 3 within 1 hour.
— If the RCS pressure safety limit is breached, the unit has to take different actions based on its current mode.
  • In modes 1 and 2, it has to restore the compliance and be in mode 3 within 1 hour;
  • In modes 3, 4, and 5, the unit is required to restore the compliance within 5 minutes.

(b) BWRX-300 operational safety limits

The OLCs in technical specifications of the BWRX-300 are based upon the safe operating envelope defined by a detailed safety evaluation. The structure of the technical specifications is based upon US NRC NUREG-1434, Standard Technical Specifications – GE BWR/6, to the extent practicable [6]. The BWRX-300 safety strategy is aligned with the IAEA SSR 2/1, Safety of Nuclear Power Plants: Design, and therefore includes a layered defense in depth approach to define mitigating features for postulated event sequences including assignment of the associated safety classes commensurate with these layers of SSCs [7].

(c) KLT-40S operational safety limits

The KLT-40S design provides safety systems to fulfil the following basic safety functions: emergency shutdown of the reactor; removal of residual heat; emergency reactor cool-down; and localization of radioactive substances within defined boundaries. The selection of parameters, values of actuation of safety systems, warning and emergency alarms were made taking into account timely detection of the beginning of the emergency process, prevention of its development, exclusion of exceeding the permissible ranges of reactor power and its doubling time, primary coolant pressure and temperature, limitation of radioactive substances release in case of an accident associated with depressurization of the primary circuit.

In accordance with the safety systems above, parameters are defined, and safe operating limits are set for them, deviations that may lead to an accident from. The five (5) parameters of safe operating limits include reactor power, power doubling period, pressure in the RCS, primary coolant flow rate that is influenced by RCP speed, and feed water flow rate in the secondary circuit system.

Safe operation limits are defined such as followings:

— The reactor power is divided into following six (6) conditions.
  - (i) above and equal to 120%N_{nom} (rated electric power capacity), at power limitation level of 100% N_{nom};
  - (ii) above or equal to 90%N_{nom} at power limitation level of 75%N_{nom};
  - (iii) above or equal to 60% N_{nom} at power limitation level of 50% N_{nom};
  - (iv) above or equal to 40% N_{nom} at power limitation level of 33%N_{nom};
  - (v) above or equal to 33% N_{nom} at power limitation level of 25% N_{nom};
  - (vi) above or equal to 24% N_{nom} at power limitation level of 10% and 15% N_{nom}.
— Power doubling period is set at the 15 seconds.
— Plant control system pressure setpoints under five (5) different core outlet temperatures were established as follows.
  - (i) greater than or equal to 15.68 MPa;
  - (ii) lower than or equal to 11.27 MPa (at t_{core} outlet (Temperature at the Core Outlet) $\geq$ 295 °C);
  - (iii) lower than or equal to 9.8 MPa (at t_{core} outlet $\geq$ 285 °C);
  - (iv) lower than or equal to 7.85 MPa (at t_{core} outlet $\geq$ 200 °C);
  - (v) lower than or equal to 5.9 MPa (at t_{core} outlet < 200 °C).
— Setpoints of primary coolant flow rate influenced by RCP speed are less than or equal to three to four RCPs at high speed $\leq$ 2350 rpm and less than or equal to three to four RCPs at low speed $\leq$ 2350 rpm.
— The setpoints of feed water flow rate in the secondary coolant system are less than and equal to 17 m^3/h at t_{core} outlet $\geq$ 295 °C and Preset feed water flow rate > 15% rated feed water flow rate.

(d) Rolls Royce SMR operational safety limits

Rolls Royce SMR has largely the same structures, systems, and components as a conventional PWR, except for being boron free during normal operation and using potassium based chemistry. The OLCs are therefore also largely similar to conventional PWRs. However, it is expected that improved analysis techniques will remove potential pessimisms, for example providing more time for corrective actions.

(e) SMART100 operational safety limits

The purpose of the technical specifications of the SMART100 is to assure the safe operation of the power plants, prevention of public disaster, and the preservation of the environment by setting basic rules for the safety of SMART100. The SMART100 has specific safety limits outlined in its technical specifications to ensure safe and efficient operation. These safety limits are categorized into the following two main areas: (i) reactor core safety limits and (ii) RCS pressure safety limit. Reactor core safety limits pertain to the core of the reactor where the nuclear reactions take place. In modes 1 and 2 of operation, the following conditions have to be met:

— The departure from nucleate boiling ratio hast to be maintained at or above 1.50;
— The peak linear heat rate that is adjusted for fuel rod dynamics, has to be kept at or below 622 W/cm. This limit ensures that the fuel rods do not overheat that could lead to fuel rod failure.

RCS pressure safety limit pertains to the pressure within the RCS. In modes 1, 2, 3, 4, and 5, the RCS pressure has to be maintained at or below 18.7 MPa.

In events of safety limit violations, specific protocols have to be complied as follows:

— Reactor core safety limit violations: If one or more of the reactor core safety limits are breached, immediate action is required. The situation has to be rectified and compliance with the safety standards has to be restored. Following this, the system has to transition to mode 3. All these actions need to be completed within a span of 1 hour from the moment the violation is detected.

— RCS pressure safety limit violations: If the RCS pressure safety limit is violated, the response depends on the current mode of operation.
  • In modes 1 or 2, the system has to restore compliance and transition to Mode 3 within 1 hour;
  • However, if the system is already in mode 3, 4, or 5 at the time of the violation, the response time is significantly shorter. Compliance has to be restored within 5 minutes.

These protocols ensure that any safety limit violations are addressed promptly and effectively to maintain the safe operation of the reactor.

(f) VOYGR™ operational safety limits

In the VOYGR™, reactor core safety limit and RCS pressure safety limit are the two parameters in the safety limit in the standard technical specifications. During mode 1 of operation, the reactor core safety limit has to comply with certain predefined parameters as follows:

— The critical heat flux (CHF) radiation has to be kept at or above the safety limits specified by the CHF correlations. For the correlation NSP4[2], the safety limit is 1.21, and for the correlation NSPN-1[3], the safety limit is 1.15;
— The peak fuel centreline temperature has to be maintained at a value less than or equal to the following function of burnup: $4901-(1.37\times10^{-3}\times$burnup, MWD/MTU$)°F$.

For modes 1, 2, and 3, regarding the RCS pressure safety limit, it is required that the pressure in the pressurizer is kept at or below 2420 psia (16.7 MPa).

If safety limits of the reactor core are breached, immediate actions are required. The situation needs to be rectified and compliance with the safety standards has to be restored. Following this, the system has to transition to Mode 2. All these actions need to be completed within a span of 1 hour from the moment the violation is detected.

If the RCS pressure safety limit is violated, the response depends on the current mode of operation:

— In mode 1, similar to the reactor core safety limit violation, the system has to restore compliance and transition to Mode 2 within 1 hour.
However, if the system is already in mode 2 or mode 3 at the time of the violation, the response time is significantly shorter. Compliance must be restored within 5 minutes.

3.2.2. Operational limits and conditions concerning operator consoles and interface

(a) APR1400 operator consoles and interface

The APR1400 is equipped with digitalized HMI system encompassing the control room systems and instrumentation and control (I&C) systems, reflecting modern computer technology. The safety console and the controls of the safety console identified in the technical specification have to be operable. OLCs require that three operator consoles in the MCR have to be operable. This requirement applies to Modes 1, 2, 3, and 4. In case of all required operator console inoperable, the operators have to shut down the reactor and maintain it in a safe condition using safety console in the MCR.

(b) BWRX-300 operator consoles and interface

The BWRX-300 safety is ensured without the need for operator actions therefore, it does not have OLCs related to the operator consoles and interface. The plant safety features to maintain the fundamental safety features

---

[2] NuScale Power Critical Heat Flux Correlation

are simple actuations that do not require continuous control. These simple actuations of passive safety and isolation features are fully automated and designed to fail into a safe state if power is lost to them. The capacity of the passive cooling systems ensures that heat removal continues in excess of seven days in the event of a station blackout and requires only simple replenishment actions thereafter. There are no post-accident monitoring type A variables[3] that are required to give the operator the needed information to perform a safety function for the BWRX-300. Because of these passive safety, fail-safe, and automated design, the operator HMI is primarily performing a defence in depth confirmation for the operator that the plant fundamental safety functions are being properly maintained.

(c) KLT-40S operator consoles and interface

The KLT-40S does not have OLCs related to the operator consoles and interface. It has its inherent and designed safety and does not need immediate operator's actions in all design based emergency conditions. Both passive and active safety systems operate without the operator's intervention for at least 1.5 hours. All passive safety systems are designed to be automatically actuated in case of loss of all AC/DC power.

(d) SMART100 operator consoles and interface

The SMART100 is equipped with fully passive safety systems as compared with most of large nuclear power plants with active safety systems. Therefore, the SMART100 can maintain a reactor in the safe shutdown condition for 72 hours without any operator action to mitigate all of design basis accidents, but the large APR1400 nuclear power plants require operator action after 30 minutes following some design basis accidents. Hence, the SMART100 adopts the operational strategy mentioned below in case of the main monitoring and control workstation (MMCW) failure instead of the limiting conditions for operation concerning operator consoles and interface in the technical specifications.

Shift operators of a SMART100 need to reside in a MCR during all operation modes, and abnormal and emergency conditions including accidents to operate the plant safely excluding MCR abandonment situation. A shift in the MCR consists of one licensed senior reactor operator (SRO), one licensed reactor operator, two assistant operators and several local operators. The large display panel, the MMCW, the SMCW and control workstation (SMCW), and the auxiliary monitoring and control workstation are installed in the MCR so that the operators can monitor and control the plant by automatic and/or manual means. The MMCW consists of an integrated workstation including the individual operator.

In case of a MMCW failure, the operators can shut down the reactor and maintain it in a safe condition using the SMCW even though emergency situations. In the event of an operator console failure in the MCR, the function of the MCR is maintained through the operational strategy for the operator console failure. The operational strategy in case of the operator console failure is as follows:

— When the SRO console fails, the SRO stays on his or her console;
— When one or more between the reactor operator, assistant operators 1 and 2 consoles fail, the reactor operator, assistant operators 1 and 2 are assigned one by one in order from the upper level console among available consoles. The operators who are not assigned move to the SMCW and perform operating assistant tasks;
— For instance, when the reactor operator console fails, the reactor operator moves to the assistant operator 1 console and the assistant operator 1 moves to the assistant operator 2 console. In addition, the assistant operator 2 moves to the SMCW;
— When the MCR is not available due to fire, toxic gas, high radiation or sabotage as well as loss of all operator consoles including the SMCW, the operators are able to move right away to the remote shutdown room (RSR) and perform adequate tasks using the remote shutdown panel.

---

[3] Type A variables that are associated with specific, planned, and manually controlled actions for which no automatic control is provided: US NRC regulation guide 1.97, Revision 5, Criteria for Accident Monitoring Information for Nuclear Power Plants (2019).

(e) VOYGRTM operator consoles and interface

The controls and indications within the VOYGR™ control room are considered non-safety related. There are backup safety related actuation switches in the MCR, but these switches are not expected to be operated during any design basis accident sequences. Post accident monitoring is not within the scope of the technical specifications since operator actions are not credited within the design, therefore no post-accident variables are considered type A per US NRC regulation guide 1.97 'Criteria for Accident Monitoring Information for Nuclear Power Plants' and are not required to have associated technical specification requirements [8].

For the VOYGR™ design, technical specifications are developed in accordance with 10 CFR 50.36 [4]. The format of the technical specifications follows the format generally used in the US nuclear industry; describing a limiting condition of operation, an applicability, and a set of actions. There are two distinct differences between the VOYGR™ design and large nuclear power plant requirements. First, the VOYGR™ design does not rely on safety related electrical power since all safety systems are comprised of valves and each valve fails to its safety position upon loss of power. Therefore, no technical specifications are associated with electrical power (traditionally section 3.8 of the technical specifications in the US). Second, there are no credited operator actions within the design basis that means there is no requirement for post-accident monitoring to be included within the technical specifications as no instrument is being relied upon by the operators to take an action.

3.3. START-UP AND OPERATIONAL APPROACH

In the context of nuclear power plant operations, it is crucial to validate the integrity of the NSSS in the aggregate for every nuclear power plant including SMRs, before they begin commercial operations. This involves conducting commissioning tests both before and after the initial fuel loading in the reactor. Understanding the operational process of nuclear power plants, such as the heat-up and cool-down of the primary system, as well as the processes for increasing and decreasing reactor power, offers insight into how each nuclear power plant model functions systematically prior to refueling outages.

Particularly, when a new nuclear power plant model needs to undergo comprehensive commissioning tests, along with such operational processes, learning the experiences of conventional large nuclear power plants can prove beneficial for the successful implementation of such procedures in SMRs.

3.3.1. Commissioning test

Commissioning tests are vital steps towards ensuring the safe and efficient operation of nuclear power plants during and after their construction. It verifies the integrity and performance of the systems and components, thereby contributing to the overall safety and reliability of the nuclear facility.

(a) APR1400 approach to commissioning test

Following integral tests during commissioning performed by shift operators for example at the APR1400 are to verify its associated systems and components together. Objectives of the cold hydrostatic test is to verify the integrity of various elements within the RCS pressure boundary. These elements include components, pipes, and welded points and this verification is achieved by pressurizing the RCS to 1.25 times (222 kg/cm^2 (21.8 MPa)) of its design pressure (175.8 kg/cm^2 (17.2 MPa)). This heightened pressure is maintained for a minimum duration of 10 minutes. Following this, an inspection is carried out on the RCS at the design pressure to ensure its robustness and reliability. Main activities of CHT are to fill and vent the RCS. This involves introducing coolant into the RCS and allowing any trapped air to escape, ensuring a homogenous environment within the system. Then RCS is heated to a temperature less than 148.8°C. Subsequently, the system is pressurized, facilitating the evaluation of its performance under conditions that simulate those during its operation. In essence, the cold hydrostatic test is a rigorous assessment of the RCS's resilience and operational readiness, contributing significantly to the overall safety and efficiency of the nuclear reactor.

The secondary hydrostatic test is to ensure the structural integrity of the components and pipes on the secondary side of the steam generator (SG). This is achieved by pressurizing the SG secondary side to 1.25 times (107 kg/cm^2 (10.5 MPa)) of its design pressure (84.3 kg/cm^2 (8.3 MPa)), maintaining this pressure for a minimum of 30 minutes, and then inspecting the SG secondary side at the design pressure. The main activities involved in the secondary hydrostatic test are at first to fill the Steam Generator (SG). This is the initial step where the SG is filled with water in preparation for the test. Secondly, The RCS is heated to raise the temperature on the SG secondary side to a maximum of 87.7°C. This is performed to simulate the conditions the system will operate under.

The pre-core hot functional test (HFT) is to confirm that the plant can transition from a cold shutdown condition to a hot standby condition using the appropriate operational procedures and this test is to demonstrate the correct integrated operation of plant systems without any fuel in the reactor vessel. The key activities involved in the pre-core HFT are to raise RCS temperature and pressure to a hot standby condition (291.3°C, 158.2 kg/cm^2 (15.5 MPa)). This simulates the conditions the system will operate under during normal plant operations. After the system has been brought to a hot standby condition, the RCS temperature and pressure are then reduced to a cold shutdown condition (23.9°C–51.7°C, 30.6 kg/cm^2 (3.0 MPa). This ensures that the system can safely transition between these two states. In APR1400, a total of approximately over 50 test procedures are implemented at their respective 19 test plateaus. These tests are designed to thoroughly evaluate the system's performance under a variety of conditions.

The post-core HFT is to confirm that the plant can transition from a cold shutdown condition to a hot standby condition using the appropriate operational procedures then demonstrate the correct integrated operation of plant systems with fuel in the reactor vessel. As key activities involved in the post-core HFT, RCS temperature and pressure are increased to a hot standby condition (291.3°C, 158.2 kg/cm^2, (15.5 MPa)). This simulates the conditions the system will operate under during normal plant operations. After the system has been brought to a hot standby condition, the RCS temperature and pressure are then reduced to a cold shutdown condition (23.9°C– 51.7°C, 30.6 kg/cm^2, (3.00 MPa)). This ensures that the system can safely transition between these two states. A total of 16 test procedures are implemented at their respective 10 test plateaus. These tests are designed to thoroughly evaluate the system's performance under a variety of conditions. This test is particularly important because it involves the operation of plant systems with fuel in the reactor vessel, adding an extra layer of complexity and safety considerations.

The power ascension test is a significant procedure, Main objectives of power ascension test are to gather data on the performance of the plant in order to verify the design parameters and demonstrate compliance with the technical specifications. This ensures that the plant is operating within the parameters set out in its design and is meeting the required technical standards. The power ascension test also demonstrates with reasonable assurance that the plant is capable of withstanding anticipated operational occurrences and transients. This means that the plant has been tested and proven to be able to handle the expected variations in operation and sudden changes in conditions.

The key activities involved in the power ascension test are as follows:

— Raising the reactor power: The power of the reactor is increased from 0% to 100%. This tests the reactor's ability to operate at full capacity and ensures that it can safely and effectively handle the maximum power output;
— Implementation of test procedures: approximately 50 test procedures are implemented at their respective 17 test plateaus. These tests are designed to thoroughly evaluate the system's performance under a variety of conditions and at different power levels.

This test is important because it involves the operation of the plant at full power, adding an extra layer of complexity and safety considerations.

(b) KLT-40S approach to commissioning test

The KLT-40S has the comprehensive start-up programme that can be divided into three phases, namely:

— Prestart-up flush and testing: It includes both primary and secondary sides to check the adequacy of all previous works on systems construction and erection, including hydraulic tests in both modes and primary circuit flush in cold mode. These testing is to be achieved at the plant manufactory site;
— Physical start-up: It incorporates first criticality and all necessary tests in the mode of first criticality to obtain data on neutron physical characteristics of the reactor core and to get confirmation of the operability and correct functioning of the reactor control and protection system;
— Power start-up: It involves power ascension and testing to determine the power, temperature and barometric reactivity coefficients, as well as the compliance of physical parameters and characteristics of the power unit systems and equipment operation in steady state and transient modes to confirm their reliable and safe operation. Also, comprehensive testing of the power unit at 100% of the reactor plant capacity has to be accomplished.

The last two phases that are the physical start-up and the power start-up for the 'Akademik Lomonosov' FNPP were carried out at the fleet base in Murmansk before the FNPP was transported to Pevek.

KLT-40 reactors are used on icebreakers, while experiencing high vibration loads. The KLT-40S is seismically designed for the same peak ground acceleration of 3g as it is necessary for icebreakers since Akademik Lomonosov is not used as an icebreaker and therefore does not experience such vibration loads during its transportation that is why it does not require to retest after transportation to Pevek.

(c) SMART100 approach to commissioning test

The SMART100 has the initial test programme including tests for demonstrating operability and functionality of the comprehensive plant systems and components from the preoperational testing to the power ascension testing. When these tests are completed, the whole plant is ready for operation with its designed normal power. The initial test programme will be performed in five distinct and sequential phases:

— Phase I, preoperational testing: The objectives of the preoperational testing are to verify the adequacy of plant design, accomplishment of plant construction in accordance with the design, and the adequacy of plant operating and emergency procedures. Furthermore, the preoperational testing enables demonstration of systems and components operability and functional abilities, familiarization of plant operating, technical and maintenance personnel with plant operation, and collection of basic performance data of systems and components. Primary and secondary side hydrostatic tests are conducted in this phase;
— Phase II, pre-core HFT: Upon completion of the preoperational testing, the integrated RCS heat-up and the pre-core HFT ensures that systems and components necessary for normal plant operation, especially those systems and components required for initial fuel loading, will safely perform their functions. During this phase, the normal operating procedures will be used to bring the plant to safe shutdown and HFT conditions. The pre-core hot function testing contains a reactor coolant flow measurement test, a heat loss test, and a RCS leak rate measurement test;
— Phase III, post-core HFT: The post-core HFT is performed following initial fuel loading and prior to initial criticality to provide additional assurance that essential systems and components for normal plant operation are functioning as expected and to obtain performance data on core-related systems and components. In-core and ex-core instrumentation tests are carried out in this phase;
— Phase IV, low power physics testing: The low power physics testing is accomplished after initial criticality. The power level is kept below than point of adding heat. During this phase, low power physics tests are performed to compare actual core parameters and reactivity coefficients with design parameters;
— Phase V, power ascension testing: The power ascension testing commences after completion of the low power physics testing and is performed to increase reactor power to full power. This testing is confirmed that the facility is operated as designed steady and anticipated transient conditions. The power ascension testing includes natural circulation test, turbine trip test, unit load rejection test, loss of offsite power test, turbine bypass valves capacity test.

(d) VOYGR^{TM} approach to commissioning test

The VOYGR™ startup and commissioning programme integrates the testing requirements of other organizations and programmes with the testing requirements of the initial test programme, inspections, tests, analyses, and acceptance criteria management, oversight, closure activities, required maintenance, surveillances, and other testing to maximize efficiency and minimize the duplication of SSC testing and system manipulations:

— Inspections, tests, analyses, and acceptance criteria: The VOYGR™ inspections, tests, analyses, and acceptance criteria programme when successfully completed by the licensee, provide reasonable assurance that the facility has been constructed and will operate in conformity with the combined license.
— Initial test programme: The VOYGR™ initial test programme was developed within final safety analysis report Chapter 14, utilizing the applicable guidance found in Regulatory Guide 1.68, The initial test programme identifies the two programme phases as preoperational test phase and startup testing phase [9].
  • Phase 1 is preoperational test phase occurs before fuel loading in the NPM composed of readiness & preparations, generic testing, system integration testing, and HFT;
  • Phase 2 is startup testing phase including fuel load and pre-criticality testing, initial criticality testing, and low power testing, and power ascension testing.
— Preoperational testing: The VOYGR™ preoperational testing is used to demonstrate inspections, tests, analyses, and acceptance criteria acceptance criteria have been met and the associated design commitments satisfied. Preoperational testing provides assurance that construction and installation of equipment in the facility have been accomplished in accordance with design. The bulk of pre-operational testing is completed prior to module installation, however there is still a small amount of pre-operational testing that needs to be performed with the module installed.
— HFT: The VOYGR™ HFT is conducted to test the RCS as close to normal operating conditions as achievable using module heating system heating to ensure proper operation of systems and components under these conditions. The HFT conditions approximate design basis temperature, differential pressure, and flow conditions to the extent practicable within preoperational test limitations. The HFT is broken into the following sections.
  • Plant system preoperational test prerequisites; module heat-up prerequisites; module heat-up and pressurization; HFT, post HFT.

Commissioning of the HTR-PM had been implemented through following three (3) different stages:

— Stage A: the pre-operation task composed of 237 tests was separated into the following two (2) categories. The cold functional test mainly incorporated the primary circuit strength test, leakage rate test, and in-service inspection test. The HFT included 30 tests such as primary circuit heating dehumidification test and leakage rate test;
— Stage B: Consisting of 52 tests titled core criticality and low power experiment was divided into three categories. The B1 named first core loading and primary criticality stage had two test items including first core loading and primary criticality test. B2 titled dehumidification stage during post-core loading had performed four tests including dehumidification after core loading fuel, handling system running-in test to verify operation capability of fuel handling system. During B3 category, low power test stage, criticality test, doppler hot spot, control rods equivalent measurement and absorbing balls equivalent measurement test had been achieved. These tests aimed to ensure the operation capability of linkage between reactor and turbine therefore primary core loading, completion of turbine running, first grid connection had been accomplished;
— Stage C: entitled power test composing 95 tasks had been conducted such as single reactor power test at 15%, 30% and, 50% and dual reactor power test at 65% and 100% called the power platform integrated transient test.

3.3.2. Plant operation approach

(a) APR1400 plant operation approach

For the plant operation approach of a large conventional nuclear power plant i.e., APR1400 model, the operation from hot standby to cold shutdown in a nuclear power plant involves a series of steps and conditions. If the unit is in an initial condition of operation Mode 3 (refer to TABLE 9. MODE OF OPERATION FOR APR1400) as follows:

— The RCS is initially in a cold state (T_{cold}) at 291 °C. This condition is maintained by the steam bypass cooling system in auto-local mode;
— The RCS pressure is maintained at 158.2 kg/cm^{2}A (15.5 MPa) by the pressurizer pressure control system in auto mode.

Then the unit will reach to the following final condition of operation mode 5 after shift crew perform cool down operation performance through plant procedures:

— The RCS is cooled to less than 65°C by the shutdown cooling system, transitioning the system to a cold shutdown state;
— The RCS pressure is reduced to atmospheric pressure;
— The pressurizer level is maintained at 100%, indicating a solid condition in the RCS.

During plant cooling operation the following limitations much be maintained for the RCS system and core safety:

— The cooldown rate of the RCS has to be less than 55.6 °C/h to ensure the safety and integrity of the system;
— The cooldown rate of the Pressurizer has to be less than 111 °C/h;
— The shutdown margin, a measure of how much the reactivity of the reactor is below the critical point, has to be 6.5 %Δk/k.

This operation is crucial in ensuring the safe and efficient shutdown of a nuclear power plant. It involves careful monitoring and control of various parameters to ensure the plant is safely brought to a cold shutdown state from a hot standby condition.

The operation from cold shutdown to hot standby in an APR1400 model involves a series of steps and conditions. If the unit is in an initial condition of Mode 5 as follows:

— RCS is initially in a cold state (T_{cold}) at around 50°C. This condition is maintained by the shutdown cooling system;
— The RCS pressure is maintained at 28 kg/cm^{2}A (2.75 MPa);
— The pressurizer level is maintained at 100%, indicating a solid condition in the RCS.

Then the unit will reach to the following final condition of operation Mode 3 after shift crew perform cool down operation through plant procedures:

— The RCS is heated to 291 °C, transitioning the system to a hot standby state;
— The RCS pressure is increased to 158.2 kg/cm^{2}A (15.51 MPa).

During plant cooling operation the following limitations much be maintained for RCS system and core safety:

— The heat-up rate of the RCS has to be less than 55.6 °C/h to ensure the safety and integrity of the system;
— The shutdown margin, a measure of how much the reactivity of the reactor is below the critical point, has to be 6.5 %Δk/k in Modes 4 and 5, and 5.5 %Δk/k in Mode 3.

This operation is vital for ensuring the safe and efficient startup of an nuclear power plant. Careful monitoring and control of various parameters are essential to make sure the plant is safely brought to a hot standby state from a cold shutdown condition.

The transition from the hot standby state to the criticality state following refuelling in a nuclear power plant is a complex process that involves a sequence of steps and conditions.

In Mode 3 of APR1400, the following conditions are required:

— The RCS temperature (T_{cold}) has to be maintained at 291 °C by the standby steam bypass control system in auto-local mode;
— The RCS pressure has to be maintained at 158.2 kg/cm^2 (15.5 MPa) by the pressurizer pressure control system in auto mode.

Upon reaching the final conditions, the operation moves to operation Mode 1 where the reactor power is set at 25%. There are certain limitations to consider during this operation as follows:

— The power ascension rate is as follows: from 0% to 40% power, there is no limit; from 40% to 100% power, the rate is limited to 3 %/h after refuelling; With a fuel defect, the rate is less than 3 %/h; and from 0% to 90% power, and less than 1.5 %/h from 90% to 100% power;
— The shutdown margin has to be 5.5 % Δk/k;
— The axial shape index has to be within ± 0.27.

The procedure step is controlled by the core management team as follows:

— All shutdown/part strength control element assemblies are withdrawn, and all regulating group control element assemblies are inserted;
— The initial boron concentration is set at the critical boron concentration from the nuclear design report (with all rods out condition);
— Control element assemblies are withdrawn to all rods out condition (except for regulating group 5) while plotting 1/M;
— Boron dilution is performed to reach criticality;
— Criticality data collection is done at 5×10^{-4} %.

The operation of increasing the reactor power from 25% to 100% in a nuclear power plant involves a series of conditions and limitations. The initial condition is the operation starts in Mode 1 with the reactor power at 25% and final conditions is to terminate operation ends in mode 1 with the reactor power at 100%. There are certain limitations during this operation involving the 5.5 %Δk/k of shutdown margin and ± 0.27 of axial shape index. The power ascension rate is as follows:

— From 0% to 40% power, there is no limit;
— From 40% to 100% power, there is no limit if there is a specified core operation history;
— From 40% to 100% power, the rate is limited to 3 %/h after refuelling;

— With a fuel defect, the rate is less than 3 %/h from 0% to 90% power, and less than 1.5 %/h from 90% to 100% power.

(b) BWRX-300 plant operation approach

Since the BWRX-300 is a natural circulation boiling water reactor, there is an important transition from cold shutdown to power operation to establish the two phase steam water interfaces. When in cold shutdown or refueling mode, a safety class 3 shutdown cooling system is in service to remove decay heat. The BWRX-300 has a simple system and approach that allows for startup directly from cold shutdown, thereby not requiring additional equipment and electrical demand for pre-heating.

During the startup, the shutdown cooling system is used in a circulation mode without cooling to enhance the core recirculation. The reactor startup is conducted simply by withdrawing control rods using the Fine Motion Control Rod Drives. After reaching criticality and the power level to allow for heat-up, the reactor coolant is heated using nuclear heat. This heat-up continues and the rate of boiling increases to the point where the shutdown cooling recirculation is secured, and the core flow is driven completely by natural circulation from the differential density caused by the core boiling. The power increase is continued by withdrawing control rods in a predetermined sequence by use of the fine motion control rod drives controlled by the plant automation system. The power increase causes steam pressurization in the RPV and the connected steam lines. During the pressurization and low power period, the RPV and steam line pressure is controlled by the turbine bypass valves that discharge steam to the main condenser. The steam turbine generator is then placed into service and pressure control is transitioned automatically to the turbine control valves. Reactor water level is automatically controlled by adjusting the speed of the reactor feedwater pump. These automated controls of power, pressure and level continue through the power ascension and at the desired power level. The BWRX-300 power control continues completely by use of control rods with the reactor power determining the output level of the steam turbine generator.

The high degree of automation along with the simple operational strategy of the natural circulation of boiling water reactors allows for smooth operation at the desired power level without complex process and system coordination. The plant can then be operated at maximum power for baseload operation or vary power to meet load following needs.

(c) CAREM25 plant operation approach

CAREM25 is nuclear reactor with natural circulation and self-pressurized that may face stability phenomena that are very different from PWR or reactors with recirculation pumps. Studies show that in natural circulation nuclear reactor, with self-pressurization, under low thermodynamic quality, the flashing effect is crucial to analyze the stable performance. Dominant destabilizing mechanism are the density waves travelling through the chimney section above the reactor core. Condensation that occurs in the steam dome has a great impact in defining the stability of the reactor. Hence, the condensation initiation can be used to tune the operational point. The reactor with natural circulation for primary core heat removal has better stability performance when the system pressure is increased. The system could be pressurized without encountering instabilities if a minimum condensation rate is maintained. It is important to correctly analyze these phenomena to define the right combination of parameters to increase the reactor stability margins.

(d) KLT-40S plant operation approach

There are two units in the design of the 'Akademik Lomonosov' FNPP with KLT-40S reactor. Each unit is composed of a NSSS including a reactor with other nuclear island equipment that are located into the metal containment, turbine section, control room and safety systems. Therefore, each unit of the KLT-40S can work independently being in either power operation mode, refueling, maintenance and cold/hot shutdown modes. It can work in load following mode from zero to full power. In case of full loss of turbine load it transfers to auxiliary (house load) operation without the reactor trip.

Heat-up, reaching criticality and power ascension of FNPP can be achieved through the following procedures:

— RCS heat-up: RCS is being brought to operating conditions with 280°C and 12.7MPa along with available pressurizer safety valves and the passive residual heat removal system. After reaching required RCS parameters the hydraulic tests are to be completed. At the same time component cooling water system (CCWS) and turbine are to be isolated, turbine bypass valves turbine bypass valves are to be available, and the condenser vacuum is built. RCS heat-up is provided by the operation of RCPs and the RCS pressure maintains by pressurizer with nitrogen gas compensation;

— First criticality: Before the first criticality, the RCS has to be in the operation conditions, and meet the chemistry requirements for the RCS, as well as all engineered safety feature (ESF) and the turbine bypass system including the turbine bypass valves are to be operable. First, the shutdown control rods are to be raised from the core, then the regulating rods are to be withdrawn up to the preliminary calculated position for further criticality and getting the core criticality by dilution of the soluble neutron absorber cadmium nitrate. The generated steam in secondary side is directed to the condenser through turbine bypass valves;

— Switch to the grid: After the first criticality is reached, the turbine is pushed by opening the stop valves and supplying steam through the control valves. The turbine starts to ramp up. The ramp-up is smooth, but at some points where the turbine has increased vibration due to harmonics, the ramp-up is faster. Once the rated speed is reached, the turbine is synchronized with the grid and the generator begins to deliver power to the grid. Subsequently turbine bypass valves are to be closed;

— Power Operation: After synchronization with the grid, a step-by-step power ramp-up begins. Power is ramped up smoothly and stops are made at 25%, 50% and 75% of rated power to make tests and capture data to confirm normal operation of the unit at different levels of power. Once 100% of rated capacity is reached, comprehensive tests are performed to ensure normal operation at floating power unit capacity.

(e) SMART100 plant operation approach

A standard design plant of the SMART100 has two units. Each unit is composed of a NSSS including a reactor and its individual reactor containment, turbine building, control rooms and safety systems. Therefore, each unit of the SMART100 can be under power operation, refuelling, maintenance and standby independently. Also, since each unit is designed to enable house load operation without a reactor trip after a complete loss of turbine load as well as daily load following operation, the SMART100 plant is capable of producing electric power from zero to full power.

Heat-up, power generation and cooldown of SMART100 can be achieved through a series of six operations of the plant: (i) the auxiliary heat-up operation, (ii) core critical operation, (iii) reactor startup operation, (iv) power operation, (v) shutdown operation by feedwater, and (vi) shutdown operation by shutdown cooling function. Here, the shutdown operation by feedwater is a means for normal cooldown of SMART100. For abnormal conditions including a loss of normal feedwater, the passive residual heat removal system can bring the reactor to the entry condition for the shutdown operation by shutdown cooling function. Details of control for each operation of the SMART100 are as follows:

— Auxiliary heat-up operation: the purpose of the auxiliary heat-up operation is to bring the RCS to the hot zero power condition where the reactor coolant temperature and pressure are 312.4 °C and 15.0 MPa, respectively. For this operation, the pressurizer safety valves and low temperature overpressure valves have to be available, and filling/venting operations and leak/hydro tests need to be completed. Also, the passive residual heat removal system, component cooling water system and turbines need to be isolated, and the condenser vacuum needs to be built, and the turbine bypass valves need to be available. Reactor coolant pressure is being controlled by the CVCS. Energizing/starting of the RCPs and the pressurizer heaters enables heat-up of the RCS and a steam volume formed in a top of the pressurizer, respectively; After forming of the steam volume, pressurizer pressure can be controlled by the pressurizer heaters and pressurizer level can be maintained by the CVCS charging and letdown flow rates. However, the RCP power is not enough to heat-up the RCS with the desirable heating rate, hence, the SMART100 is equipped with the preheating system in the balance of plant side as an additional heat source. Thus, the RCS can reach the hot zero power condition with the aid of the preheating system, RCPs and the pressurizer heaters;

— Core critical operation: before the core critical operation, the RCS has to be under the hot zero power condition and meet the chemistry requirements for the RCS. Additionally, all ESF and the turbine bypass system including the turbine bypass valves have to be operable. After energizing the control rod drive mechanisms, complete withdrawal of the all shutdown control rods, control of the regulating control rods and dilution of boron can result in core critical;

— Reactor startup operation: prior to the synchronization of the turbine generator, the power of the reactor core can be progressively escalated up to 20%. The increase of 12% core power will be achieved by manipulating the regulating control rods. At the same time, the appropriate feedwater flow rate according to the core power can be also injected into steam generators by throttling the feedwater control valves and the turbine bypass valves and regulating the feedwater pumps;

— Power operation: for synchronization of the turbine generator, opening of the turbine stop valves and throttling the turbine admission valves and the turbine bypass valves are conducted. After a turbine load reaches 20%, the heat balance of the plant can be checked. Power ascension can be achieved from 20% to 100% by discrete steps, e.g., 10%;

— Shutdown operation by feedwater: the shutdown operation by feedwater can be started from the plant power of 20%. After the turbine bypass system is operable, the turbine can be isolated from the main steam lines. For a reactor trip, the shutdown control rods are completely inserted and the CVCS injects high concentration boron acid water into the RCS. At the same time, the feedwater flow rate needs to be controlled to meet the RCS cooling rate under its limitation. This feedwater flow rate may be smaller than a criterion to supress the flow instability in the steam generators. Thus, only one main steam line and one feedwater line are used to maximize the flow rate per a steam generator tube by isolating other three main steam lines and three feedwater lines. A SMART100 unit has eight steam generators, and each two steam generators are connected to a main steam line and a feedwater line. Therefore, there are four main steam lines and four feedwater lines for a SMART100 unit;

— Shutdown operation by shutdown cooling function: the shutdown operation by shutdown cooling function can be conducted after the low temperature overpressure valves are connected to the RCS and the RCS temperature is lower than the saturation temperature of operating pressure of the component cooling water system. First, cold water injected from the component cooling water system into a feedwater line cools the RCS via two steam generators. The water discharged from the steam generators returns to the component cooling water system through a main steam line. Then, the feedwater system can be isolated. If required, another feedwater line and main steam line can be additionally connected to the component cooling water system. i.e., two steam generators are on standby for the additional cooling rate of the RCS. This shutdown operation brings the RCS under the refuelling mode.

(f) VOYGR™ plant operation approach

The startup, heat-up, and power accession of the VOYGR™ design are as follows:

— Mode 5 Refuelling: VOYGR™ NPMs will need to be shut down and disassembled every 18 to 24 months to allow for refuelling, load shuffle, required inspections, and maintenance. Following the completion of the refuelling, lower containment, upper module, and operating bay work windows the NPM is assembled in the reactor flange tool and containment flange tool respectively. When the last reactor flange bolt is fully tightening the NPM enters Mode 4.

— Mode 4 Transition: at this time the NPM is in a flooded state. Both the reactor and containment are filled to the same level with emergency core cooling system valves open with borated water at the concentration of the reactor pool. Once tension of the closure nuts has been verified the NPM is transported via the reactor building crane to its associated operating bay within the reactor pool.

- Operating bay: once the NPM is secured in the operating bay maintenance craft connect module instrumentation, controls, hydraulics, power, piping, manway covers, assemble installation, reposition neutron monitors, and complete regulatory required leak rate testing. Once the module is connected the following startup prerequisites are verified.
  - systems aligned for initial start;
  - liquid rad waste is ready to accept letdown;
  - initiate condensate and feed system;
  - all support systems are available.
- Transition to Mode 3 and heat up activities: once the startup prerequisites have been verified the operator will transition the module to Mode 3 and start the non-critical heat up by performing the following.
  - Verify Mode 3 Technical Specifications surveillance are complete;
  - Reset containment isolation signal and CVCS trips signals;
  - Pressurize the pressurizer with nitrogen gas (N2);
  - Once the CVCS is uninsulated the module is considered in Mode 3;
  - Place the CVCS in operation to start initial clean-up of the reactor coolant through demineralizers and allows for the adjustment of RCS level and boron through the combination of demineralized water or boron injection and letdown;
  - Start an auxiliary boiler;
  - Operators close the emergency core cooling valves to isolate the RCS from the containment and allow for heat-up activities to commence;
  - Drain the containment.
- Transition to Mode 2 hot shutdown: once reactor coolant level is adjusted within the pressurizer normal operating band, the pressurizer heaters are energized to create a steam space and raise reactor coolant pressure. Operators will initiate module heating, perform Mode 2 surveillances, commence warmup of the steam piping, and perform required rod testing and when RCS temperature reaches 174°C (345 °F) enter Mode 2.
- Reactor critical approach: the VOYGR™ design can achieve a critical condition using either dilution or control rods, but current design preference is to adjust the boron concentration and then perform a critical approach using control rods only. An estimated critical rod position is calculated based on stable reactor coolant temperature, core poison condition (such as Xenon), and boron concentration. Operators will perform the following.
  - Dilute reactor coolant to target concentration;
  - Perform Mode 1 surveillances;
  - Verify chemistry is within limits;
  - Withdrawal shutdown bank of control rods;
  - Enter Mode 1 (Mode 1 is entered administratively with all shutdown rods withdrawn);
  - Withdrawal regulating rods to criticality.
- Critical: once critical, operators will perform any required low power physics testing to verify core conditions and then heat up the reactor coolant to normal operating temperature using nuclear heat (the CVC module heating system is secured when no longer needed). When sufficient heat is being generated, the main steam system is warmed up and associated systems placed in service to support eventual turbine synchronization (such as steam seals, lube oil, vacuum in main condenser, etc.).
- Turbine synchronization: reactor power is raised using the turbine bypass valve to a power level of approximately 20%. Turbine warmup, spinning to operational speed, and preparations to place in service are then completed. The turbine generator is then electrically synchronized to the grid to begin producing electricity. The turbine bypass valve is slowly closed in a controlled fashion as the main turbine is then able to pick up additional load.
- Power accession to 100%: once the turbine bypass valve is fully closed, reactor power is then adjusted using the feed system to raise feed flow, control rods to maintain reactor temperature, and the turbine governor valve to admit additional steam to the main turbine. At full reactor power the control rods will be fully withdrawn and additional changes to reactivity to compensate for Xenon would be performed using boration or dilution through the CVC system as necessary.

3.4. OPERATING PROCEDURES AND GUIDELINES

Operating procedures in nuclear power plants are essential for ensuring safety, efficiency, and regulatory compliance. Operator's procedures among operating procedures play a crucial role guiding the operation, maintenance, and emergency response, among other aspects of a nuclear power plant. Learning shift operator's procedures in nuclear power plants will let us know plant operation well.

(a) APR1400 operating procedures

In APR1400 models, computerized procedures systems are utilized as advanced systems designed to enhance the safety and efficiency of plant operations. Here are some key aspects. Computerized procedures systems can automate certain tasks, reducing the potential for human error and improving the overall efficiency of plant operations. Computerized procedures systems provide crucial support to operators, helping them manage complex tasks and make informed decisions. By providing clear, step-by-step instructions for various procedures, Computerized procedures systems can help enhance the safety of nuclear power plant. Computerized procedures systems are designed to improve human performance by providing clear, concise, and accurate procedural guidance.

Normal operating procedures are to guide the routine operation of the plant, including starting and stopping equipment, changing modes of operation, and performing regular maintenance. Abnormal operating procedures are used when the plant deviates from normal operations but is not in a state of emergency. They guide the operator in returning the plant to normal operating conditions. During serious incidents or accidents, emergency operating procedures are utilized. They provide instructions for mitigating the consequences of the event and maintaining or restoring the safety functions of the plant.

(b) BWRX-300 operating procedures

The BWRX-300 operational control includes a high degree of automation in addition to the simple fail-safe isolation and passive features. These automated sequences include interfaces for the operator to periodically confirm readiness to move to the next evolution beyond the hold point. In order to allow for this type of automated control with periodic confirmation at hold points, the operating procedures are planned to be largely electronic and embedded into the plant controls with indication on the HMI.

(c) SMART100 operating procedures

Preliminary normal and emergency operating procedures for the standard design of the SMART100 have been developed considering operating experiences of large nuclear power plants and systems and components introduced in the SMART100. Computer based normal and emergency operating procedures for the SMART100 are planned to be submitted in a phase for construction permission and operating license.

(d) VOYGR™ operating procedures

Normal and emergency operating procedures for the VOYGR™ design are embedded within the non-safety control system and are considered type 3 computer-based procedures per IEEE-1786 'Guide for Human Factors Application of Computer Based Operating Procedures (COPS) at Nuclear Power Generating Stations and other Nuclear Facilities'. Type 3 procedures allow for direct indications to be embedded within procedure steps as well as providing soft controls for certain components. For example, a procedure may direct start a pump at a prescribed suction pressure. The suction pressure may be displayed dynamically within the procedure and the pump control may also be included within the procedure step. The Intent is to limit the potential mis-operation of components between multiple units. Figure 9. shows the example type 3 procedure with embedded indication and soft control of the VOYGR™.

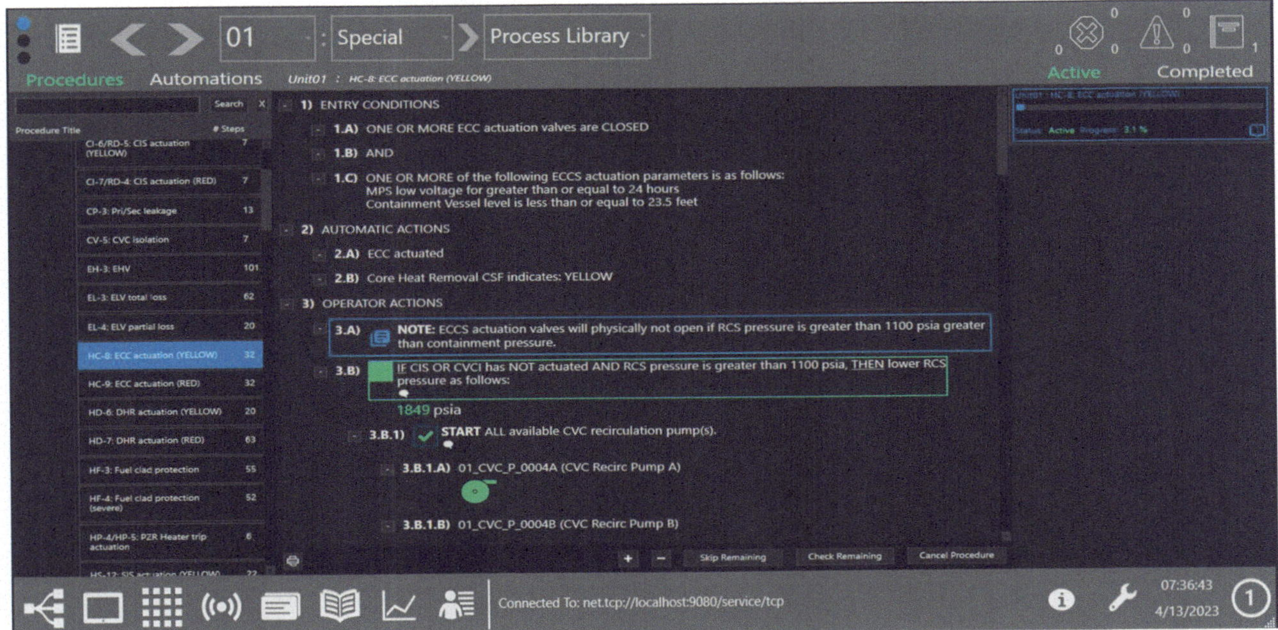

FIG. 9. Computer based operating procedure with embedded indication and soft control of VOYGR™. (Source: NuScale Power Inc., United States, with permission).

4. CONTROL AND MANOEUVRE APPROACH

4.1. CONTROL OF NUCLEAR STEAM SUPPLY SYSTEM AND PLANT CONTROL SYSTEMS

The NSSS is a crucial component of a nuclear power plant. It is responsible for producing steam to supply the turbine generator units that in turn generate electricity. NSSS process control system regulates the control systems of NSSS and plant control system is an integrated control system that provides reactor power level control. Therefore, NSSS process control system and plant control system perform crucial roles in operating nuclear power plants safely and efficiently.

(a) APR1400 control and maneuver approach

In APR1400, The NSSS process control system consists of the pressurizer pressure control system, pressurizer level control system, feedwater control system, steam bypass control system, boron dilution alarm system, and single control loops of the CVCS. The plant control system includes the reactor regulating system, reactor power cutback system, and digital rod control system.

(b) BWRX-300 control and maneuver approach

The BWRX-300 is a direct cycle boiling water reactor leveraging the simple approach of promoting and allowing boiling directly in the reactor core. The steam that is developed in the core transitions through a chimney region above the core and then exits through steam separators and steam dryers thereby providing dry steam to the steam turbine generator. The turbine control valves automatically control the reactor pressure by throttling appropriately that also controls the steam turbine generator output. The reactor water level is automatically controlled by adjusting the reactor feedwater pump speed. When the reactor power level is adjusted by the control rods as demanded by the plant automation system for load following or other reasons, these other control parameters for water level and pressure respond accordingly to maintain these in their proper band.

The BWRX-300 is a natural circulation reactor. The reactor recirculation flow rate varies with reactor power and the associated boiling rate. The reactor power is directly controlled by control rod position changes based upon demands from the plant automation system. As the power is increased, the boiling rate in the core increases and the reactor core flow rises due to the density head difference such that the appropriate power to flow ratio is maintained naturally without the need for any control signal associated directly with flow. When power is reduced, the reverse occurs to cause an associated flow reduction. The BWRX-300 is a direct cycle boiling water reactor and therefore the main steam and feedwater system operation is integral with the reactor operation. The main steam is supplied directly from the RPV, and the feedwater is returned directly to the RPV and is used for RPV water level control.

(c) KLT-40S control and maneuver approach

In KLT-40S, the load following mode relies on the inherent safety of the reactor due to negative reactivity coefficients. The temperature of the primary coolant is monitored, and reactivity control is implemented by changes in the position of the control rods and the concentration of cadmium nitrate in the primary coolant. During the load following process, both automation and operators are involved, that regulate reactivity, feedwater flow rate, and turbine bypass flow rate. The KLT-40S has a reactor control and protection system that ensures the NSSS safety of in all operation modes. Small load changes are compensated by moving the control rods, in case of necessity of significant load change of the NSSS load, water exchange with change of dissolved absorber concentration in RCS is performed.

Movement of rods is controlled by reactor control and protection system similarly to the way it is organized in large nuclear power plants, temperature, pressure and level in RCS are maintained by appropriate regulators. Appropriate safety systems, both active and passive, are provided in case of violation of the NSSS operation limits and conditions.

(d) SMART100 control and maneuver approach

The reactivity feedback characteristic of the NSSS of SMART100 is designed to follow the turbine load. For load following, reactor coolant temperature is a control parameter and is adjusted by changes in total reactivity as implemented by the control rod assembly position changes and boric acid concentration changes in the primary coolant. The capability of the NSSS to follow turbine load depends on the ability of the control systems or operator to adjust reactivity, feedwater flow, steam bypass flow, reactor coolant inventory and an amount of energy in the pressurizer within normal operating limits of the NSSS.

SMART100 has a power control system and a NSSS process control system that automatically control major systems and/or components of the plant, but they are not essential for safety. The power control system consists of reactor regulating system, control rod drive mechanism control system and reactor power cutback system. The NSSS process control system is composed of feedwater control system, steam bypass control system, pressurizer pressure control system and pressurizer level control system. These systems utilize digital processing equipment.

The reactor regulating system receives signals of reactor power, turbine load index and primary coolant temperature and transfers signals for insertion or withdrawal of control rods to the control rod drive mechanism control system that controls the CRDM coils to move the control rods. The reactor power cutback system is a control system designed to accommodate certain imbalances by reducing reactor power rapidly. The rapid decrease in reactor power can be obtained by providing high speed insertion of the control rods and reducing feedwater flow rapidly. The feedwater control system controls feedwater control valves and feedwater pumps to provide the appropriate feedwater flow into steam generators and to adjust main steam pressure from the steam generators. The steam bypass control system is designed to control turbine bypass valves to increase plant availability by removing excessive NSSS stored energy. The pressurizer pressure control system and pressurizer level control system are designed similarly to those in commercial large nuclear power plants to maintain the pressurizer pressure and level, respectively.

(e) VOYGR^{TM} control and maneuver approach

To change reactor power in the VOYGR™ design when the turbine generator is synchronized to the grid, the operator initiates a power change through an automation interface and sets a target power and target rate of change. Once initiated, three separate controllers will be used to create the desired change. The power change is driven by a change in the speed of the variable speed main feed pumps. To raise power, the speed of the feed pumps will rise at a rate set by the operator for the target power change rate. As the feed pump speed rises, more feed flow is admitted into the steam generators. This additional feed flow results in additional steam generation. The superheated steam at the outlet of the steam generator raises the steam pressure within the main steam system. To counter this rise in steam pressure, the main turbine governor valves automatically start to open to maintain a constant pressure. This has the effect of allowing more steam to the main turbine and therefore increasing the electrical output of the generator. The other effect of raising the feed flow is that reactor coolant temperature will tend to lower. To counter this effect, control rods automatically will withdraw to maintain the same coolant temperature. The combined effect of these three actions is a rise in reactor power and a rise in the electrical output of the generator. The feed pump speed will stop rising once the desired reactor power value is obtained. The automation continues to monitor reactor power and maintains all the controllers in automatic until the effects of the feed heating system are complete.

4.2. FLEXIBLE OPERATION

Due to growing renewable energy sources, there has been a significant rise in the demand for flexible operation of nuclear power plants. As a result, most SMRs are equipped with load following and frequency control operation. Traditional large nuclear power plants including those in the EDF fleet have successfully demonstrated flexible operation throughout the year without significant issues. Therefore, understanding the safe practices of how large nuclear power plants conduct flexible operation can serve as valuable sources for upcoming near term SMRs.

4.2.1. Operational approach for flexible operation of conventional nuclear power plants

The EDF operates a fleet of 56 nuclear reactors providing flexible and low carbon generation. Since the 1980s, the EDF nuclear fleet has been providing 70 to 75% of France's generation along with mainly hydropower. This has resulted in very low carbon generation (6 g CO_2/kWh life cycle). Since 2010, the renewable energy generation increase has amplified the variation demand that requires more need of flexibility for nuclear power plants.

Most nuclear power plants are currently optimized for base load operation. With the growth of intermittent renewables, grid operators encourage nuclear utilities to operate more flexibly. Flexibility of nuclear plants provides permanent and reliable electricity generation, while allowing development of renewables and keeping carbon emissions low.

In case of high renewable generation, if a nuclear plant is not flexible, it will stop and restart daily or consider final shutdown, whereas flexible nuclear reactors can reduce power and go back to full power when required. Flexible reactors can accommodate potential overcapacity on electricity markets. When renewables are available and electricity demand is lower, reducing nuclear power to save fuel, available later when the market will need, is cost effective.

To reduce nuclear generation, EDF nuclear power plants have a large load variation operation mode. This large load variation operation mode is validated by the regulator and agreed with the grid operator. French nuclear reactors can reduce their power twice a day, ramp-up/ramp-down within 30 minutes between 100% and 20% nominal power, to regulate services to the grid. These large load variation operations are performed through control rod moves (quick) and boric acid concentration changes (low).

In addition, EDF nuclear power plants guarantee grid frequency by automatic load variations. Frequency inertia and voltage control are two important factors that are considered in this operation. The secondary frequency control operates over a longer timeframe of up to 15 minutes. This is to achieve the 'frequency restoration reserve', an operational reserve activated to restore grid frequency to the nominal frequency at national and European scales.

In 2014, an EDF study identified a 20 GW variation capacity need by 2020. In March 2019, there was a record of variation of 19.5 GW successfully passing through. There is no need to increase each reactor capabilities, but more reactors are requested simultaneously to decrease and increase power.

EDF flexible reactors are allowed by the regulator to operate in load following model. All power variations remain within the technical specification domain where safety has been demonstrated. As a result, fuel thermodynamic characteristics stay beyond technical specifications limits proven to be hazardless of fuel integrity. Flexible operation is part of control room operator's training and qualification. Full scope simulators are used for such a training. The technical specifications and procedures give general instructions and advanced computerized tools help to forecast the boric acid strategy. There is no noticeable radioactive release into the environment in terms of liquid, gaseous or solid wastes. In addition, no noticeable impact on chemistry or radiological condition for workers. There is also no impact on the plant lifetime and the additional corrective maintenance during outage is unnoticeable. And the relation between unplanned outages and flexibility is small and hardly noticeable as a statistical correlation.

Nuclear reactors can be operated in flexible mode, it is safe and cost-effective. Other flexibility sources to have been used such as interconnections, hydroelectric power, and batteries. Nuclear and renewables can build together a low carbon generation mix.

4.2.2. Operational approach for flexible operation of multi-module small modular reactors

(a) SMART100 flexible operation

The load following operation of the SMART100 is simpler than that of large nuclear power plants because only a single bank movement (insertion and withdrawal) is required to respond to small reactivity change. This feature minimizes change in reactor coolant temperature. Relatively high lead bank worth is needed due to a small number of fuel assemblies. The shorter core height leads to rapid damping of the xenon oscillation. The daily load following performance simulation of the SMART100 core shows that radial peaking factor, 3D peaking factor and the axial offset were satisfied within design limit.

The standard design plant of the SMART100 has two units, but each unit is equipped with its individual turbine generator. Thus, each unit can follow the load demand separately. Design characteristics of each unit of the SMART100 for load following operation are as follows:

— Step power change of ±10% in the range between 20% and 100% power;
— Ramp power change of ±5%/minute in the range between 20% and 100% power;
— Daily load following operation of 100%-50%-100% power over 90% of the cycle life;
— Transition to house load operation without a reactor trip in case of a complete loss of load from rated power.

(b) VOYGR™ flexible operation

VOYGR™ design offers flexible power operations defined as any mode of plant operation where electric power output is varied in response to regional electrical grid demands. The VOYGR™ has features enhancing its ability to load follow such as the Table 15. below that include:

— Module dispatch: taking one or more modules offline for extended periods of low grid demand or sustained wind output;
— Reactor power change: changing power levels at one or more modules to compensate for hourly changes in demand or wind generation. Reactor power can be adjusted between 100% to 50% using control rods;
— Turbine bypass: bypassing turbine steam to the condenser for one or more modules.

TABLE 15. LOAD FOLLOW METHODS OF VOYGR™

Methods	Power increase	Power decrease	Assumptions
Module dispatch	Hot shutdown → 100% (13 h)	100% → Hot shutdown (30 min)	• Refuel schedule • Increase rate at fuel design limit: 50%/h • Decrease rate: 200%/h
Reactor power change	50% → 100% (60 min)	100% → 50% (15 min)	• Increase rate at fuel design limit: 50%/h • Decrease rate: 200%/h
Turbine bypass	20% → 100% (27 min)	100% → 20% (~8 min)	• Increase rate at turbine load limit: 3%/min • Decrease rate: 10%/min

Table 16. also presents comparison of VOYGR™ capability to utility requirements based on Electric Power Research Institute utility requirements document (Revision 13) [10].

TABLE 16. COMPARISON OF VOYGR™ CAPABILITY TO UTILITY REQUIREMENTS

EPRI utility requirements document (URD), Rev.13	VOYGR™ capability	Licensing basis
24 hour load cycle of 100% → 50% → 100%	Exceeds (100% → 20~50% → 100%	Requires licensing topical report (LTR) submittal
Ramp rate of 25%/h	Exceeds (50%/h)	SDA/COLA
Capable of automatic frequency response	Meets	SDA/COLA
Step change of 20% in 10 min for grid stability	Meets	SDA/COLA
Frequency variation tolerance	Meets	SDA/COLA

An example of VOYGR™ load following to compensate for generation from the Horse Butte wind farm and daily demand variation such that NuScale design as shown in Figure 10. meets or exceeds Electric Power Research Institute utility requirements document (Revision 13), load following and other ancillary service requirement. As a result, flexible operations of VOYGR™ enable further growth in renewables.

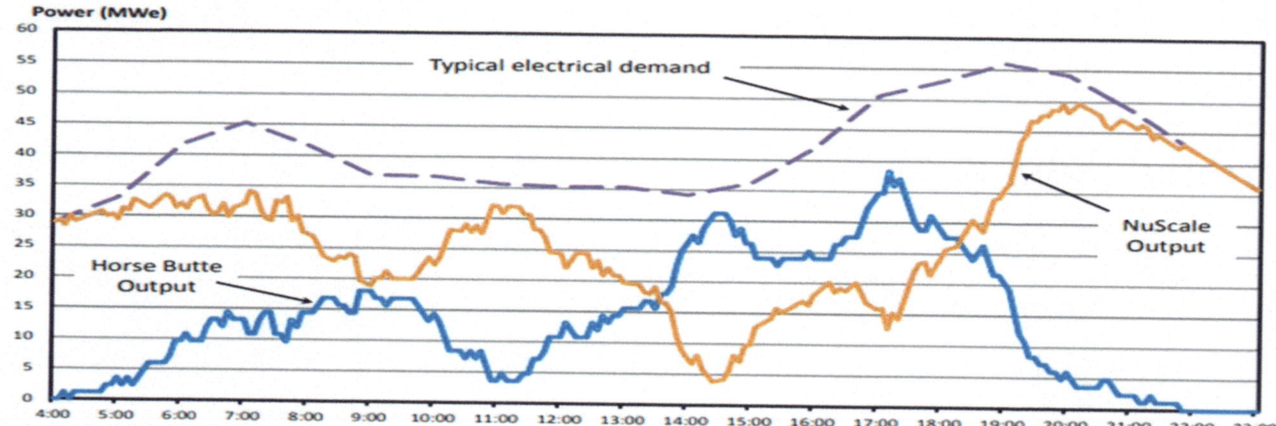

FIG. 10. Load following of VOYGR™ with the Horse Butte wind farm and daily demand variation. (Source: NuScale Power Inc., United States, with permission).

As two load following options are presented in Fig. 11, the upper graph uses only turbine bypass, and the lower graph shows a combination of reactor power manoeuvring and turbine bypass.

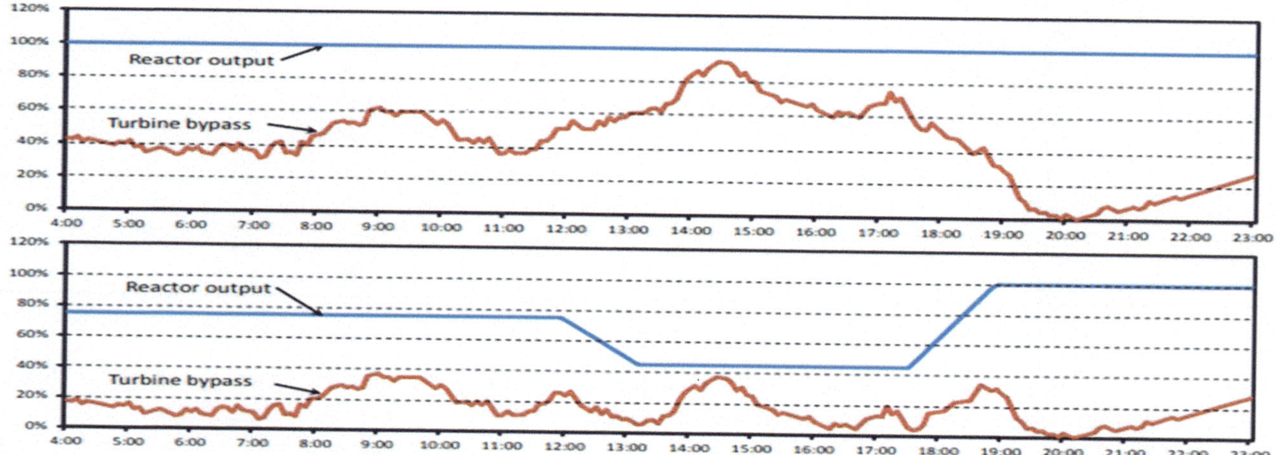

FIG. 11. Two load following options of VOYGR™. (Source: NuScale Power Inc., United States, with permission).

5. FUEL INTEGRITY OF SMALL MODULAR REACTORS

5.1. GENERAL FUEL INTEGRITY IN SMALL MODULAR REACTORS

Fuels in land based water cooled SMRs and high temperature gas cooled SMRs have been selected due to the depth of understanding and the notable progress made in their deployment. Those SMRs set it apart from traditional power generation methods. SMRs utilize a condensed version of conventional pressurized water reactor (PWR) fuels that allows for a more compact and efficient design.

In near term aspect of SMR fuels, use of high assay low enriched uranium or low enriched uranium plus and accident tolerant fuels are promising, but there are also significant challenges that need to be overcome to ensure their safe and efficient operation. The use of high assay low enriched uranium or low enriched uranium plus is anticipated in the near future. This would allow for a higher burnup and longer refuelling cycle, enhancing the efficiency of the reactor. The compact design of these reactors offers the advantage of site flexibility, allowing them to be installed in a variety of locations and the economic viability of these reactors is a promising prospect. Their efficient operation and longer refuelling cycles could lead to significant cost savings.

However, there are still unresolved issues related to high burnup operation, especially at very high burnups that exceed regulatory review limits. These include the formation of a high burnup structure in the pellet periphery that could affect the fuel performance. The potential for a burst release of fission gases during high burnup operation is a significant concern. This could pose a risk to the safe operation of the reactor. Fuel fragmentation, relocation, and dispersal (FFRD) during a loss of coolant accident is another challenge. This could potentially lead to fuel damage and compromise the safety of the reactor. The possibility of rod ejection during a reactivity initiated accident is a safety concern that needs to be addressed.

Accident tolerant fuels are designed to withstand accidents better than traditional fuels, providing an added layer of safety and security. The use of accident tolerant fuels can lead to a significant improvement in fuel reliability during the operational states of the plant, ensuring smoother and more efficient operations. accident tolerant fuels can also facilitate the approach to high burnup operation, allowing for more energy to be extracted from the fuel. However, the path forward is not without its challenges. While accident tolerant fuels can provide a limited increase in the time available to respond to an accident, there is a need for more revolutionary concepts to significantly extend this coping time. The use of accident tolerant fuels can come with an economic penalty, as these advanced fuels can be more expensive than traditional nuclear fuels.

In boron and lithium hydroxide (LiOH) usage in nuclear reactors, boron is used as a neutron absorber in the RCS to regulate fission reactions in nuclear reactors. Depending on the specific design of the reactor, it may eliminate the use of boron and LiOH in its chemistry that can reduce the chemical drawbacks associated with boron use in the primary systems. One of the key features of this SMR is its ability to operate the fuel assembly under natural circulation and at low flow rates. This feature enhances the safety profile of the reactor by reducing reliance on active mechanical systems for coolant circulation.

However, like all technologies, the water cooled SMR is not without its challenges. The mechanical design verification process can be complex, given the novel nature of its components. The reactor's power manoeuvring operation can be affected by a large axial power offset and rapid power changes caused by mechanical control systems, requiring careful management and control.

One of SMR operations concerns is the potential thermal hydraulics impact on fuel integrity, such as the departure from nucleate boiling ratio. This could potentially lead to fuel damage if not properly managed. SMRs may face a loss of economic efficiency due to a reduced discharge burnup. This means that the reactor may not be able to extract as much energy from the nuclear fuel as possible before it needs to be replaced that could increase operational costs.

For the shortcomings of SMR fuel for land based water cooled modules, mechanical design verification is a challenge due to the complex geometry of SMR fuel assemblies. Fuel performance data to support desirable or unique features can also be a challenge due to the limited operating experience of SMRs including boron free and/or Li free operations. Regarding control rod ejection accidents and pellet cladding interaction (PCI), they are likely to be more limiting than for conventional plants due to the small size of SMRs and their fuel assemblies. Power manoeuvring operation can also be challenging due to a large axial power offset and a rapid power change caused by mechanical control systems.

For high temperature gas cooled SMR fuel, the typical design is Tristructural-Isotropic (TRISO) coated particle with a fuel kernel surrounded by several coating layers. A prismatic high temperature gas cooled reactor has coated particles contained in a fuel compact inserted into a graphite block as fuel elements. The pebble bed high temperature gas cooled reactor has coated particle fuels (enrichment: 8.5) embedded in a diameter 60 mm graphite matrix as spheres fuel. TRISO can also be used for water reactors, gas reactors, and molten salt reactors.

High temperature gas cooled SMRs operate at very high temperatures and use gas as a coolant. They typically use TRISO coated particle fuel. Liquid metal cooled SMRs use liquid metal as a coolant and are known for their high efficiency. They use mixed oxide or metallic uranium-zirconium (U-Zr) fuel when sodium cooled and oxide or nitride fuel when lead or lead-bismuth cooled. Molten salt type SMRs use molten salts as both a coolant and a means of fuel transport. micro modular reactors (MMRs) are smaller, more flexible reactors that use fully ceramic microencapsulated (FCM) fuel. They are designed for remote locations and smaller grids. Each of these reactor and fuel types has its own unique advantages and challenges, and they represent the diverse approaches being taken to advance nuclear technology. The limitations of high temperature gas cooled SMR fuel, such as restricted packing fraction (Uranium loading in fuel element), managing graphite powder for reactor use, and a substantial amount of spent coated particle fuels, can pose difficulties. Deployment challenges include the need for verification through in-pile irradiation tests up to a significant discharge burnup. The application of 'advanced' fuel manufacturing methods like additive manufacturing related to 3D printing can also be problematic due to the small size of TRISO.

The IAEA has proposed deliverables from meetings including completion of TECDOC on up-to-date information collected since 2010 covering assessment of the effect of variations in the fuel fabrication parameters (sintering temperature, etc.) and methods (for kernel fabrication, coating process) on the as-fabricated fuel characteristics (microstructure, porosity, etc.) and hence on fuel performance. Assessment of radioactive waste associated with the graphite matrix embedding TRISO particles for gas cooled reactors is also included.

5.2. FUEL INTEGRITY DURING FLEXIBLE OPERATION IN SMALL MODULAR REACTORS

One of key driving forces of SMR development is fulfilling the need for flexible operation, reducing capital costs, and enhancing safety and security. Flexible Operation refers to the ability of a nuclear power plant to adjust its electrical output to match the electrical demand and control the frequency of the electrical system.

No fuel failures occur if fuel rods are operated within the respective domain of technical specifications. Margins to failure need to be quantified to ensure fuel reliability under flexible operation conditions. With quantified margins, operators can relax some of the constraints imposed on reactor operation to better accommodate power grid requirements. Margin assessment is of interest for anticipated operational occurrences following extended low power operation.

Flexible operation and related power changes of nuclear power plants usually involve power variations, and the high stress on the fuel cladding upon a power increase, via a strong PCI that may impact fuel integrity through effects on the thermal mechanical performance of the fuel rod such as pellet cladding interaction/stress corrosion cracking (PCI/SCC). Failure risk of PCI/SCC needs to be assessed by considering additional anticipated operational occurrences related to the flexible operation. This requires a good understanding of PCI/SCC mechanisms, extensive experimental investigations, validated calculation tools, and qualified PCI/SCC design verification methods.

IAEA TECDOC 1860 provides a review of up-to-date progress on international fuel community's R&D efforts to address the challenges above, and a critical review of state-of-the-art knowledge of PCI/SCC to share information on advanced experiments, modelling, fuel design methodologies, and operating guidelines for the flexible operation of nuclear power plants [11]. PCI/SCC needs to be taken care of during reactor system operation. It reflects fuel behaviour during power maneuvering. Better understanding is needed on some basic phenomena of PCI/SCC. To achieve more economical and flexible operating conditions, continued experimental and analytical work is necessary.

5.3. FUEL CHARACTERISTICS IN SMALL MODULAR REACTORS

(a) KLT-40S

The KLT-40S uses cermet fuel that means uranium dioxide granules (UO_2 volume fraction less than 70%) arranged in a metal matrix (silumin alloy) with U235 enrichment below 20% (low enriched uranium in the IAEA terms). Such fuel is characterized by the absence of direct contact between fuel particles due to their uniform distribution in the metal matrix. This is achieved by using spherical fuel particles pre-coated with matrix material and their isostatic pressing into the fuel element cores. Cermet fuel can compensate for 'solid' swelling of the fuel element core and localize 90% of fission products in the UO_2 granules.

(b) Rolls Royce

The Rolls Royce SMR uses conventional UO_2 fuel and is designed to operate within recommended limits. They are using advanced methods for uncertainty quantification to confirm that the operational characteristics do not breach any limits. The reliability of systems and components is assessed from a safety and availability perspective, and availability targets have been set on the design, with a reliability centred maintenance approach determining the appropriate maintenance activities and scheduling to support the required availability and reliability targets.

(c) SMART100

The SMART100 adopts the fuel assembly consisting of a 17x17 square array of 264 fuel rods with an active length of 2.0 meters. The fuel for the SMART100 is less than 5% enriched UO_2 in the form of ceramic pellets and is encapsulated in pre-pressurized fuel rods that form a hermetic enclosure. The integrity and reliability of the fuel assembly has been already used and proven in many large PWRs in Republic of Korea. The fuel performance tests, and the critical heat flux measurement tests were conducted for the standard design of the SMART100.

(d) VOYGR™

The VOYGR™ fuel assembly design is already in use in large LWRs with more than 20,000 fuel assemblies delivered. NuFuel-HTP2™ is a shorter version (1.83 m heated length) of the existing Framatome HTP™ fuel design (3.66 m heated length). Each component such as grids, nozzles, fuel rods, pellet, and cladding in the NuFuel-HTP2™ assembly is based on the existing Framatome 17x17 HTP™ fuel assembly design. Framatome 17x17 HTP™ assemblies have been deployed in 52 reactors supporting power generation in 11 countries. M5® cladding is used in the NuFuel-HTP2™. More than five million M5® clad fuel rods have been used in 84 reactors around the globe. Several prototypic NuFuel HTP2™ fuel assemblies were manufactured and Framatome's standard suite of mechanical and hydraulic testing was conducted to characterize the fuel assembly to support US NRC licensing submittal.

6. OPERATING PERSONNEL MANAGEMENT AND HUMAN MACHINE INTERFACE

6.1. ARRANGEMENT OF MAIN CONTROL ROOM CREW

The MCR crew in both large nuclear power plants and SMRs play a crucial role in the safe and efficient operation of the plant. However, the number of operators and their arrangement in an MCR can vary among large nuclear power plants and SMRs. In large nuclear power plants, the MCR crew typically includes a shift supervisor, a reactor operator, and one or more assistant reactor operators. In the case of SMRs, the number of operators can be greatly reduced due to the plant's smaller size and increased automation. The control strategy of SMRs generally goes beyond the traditional rules for control rooms, so the licensees need to demonstrate the safe operation of the plant with that configuration.

(a) APR1400 crew arrangement for main control room

At a APR1400 plant in Republic of Korea as a model of conventional large nuclear power plants, each unit installed in its individual reactor containment and auxiliary building including a MCR is connected to its individual turbine generator. The two units share a common compound building. Each MCR in APR1400 plants typically has a standard crew of five per shift including the following positions: shift supervisor (SS), shift technical advisor (STA), reactor operator, turbine operator, and electrical operator. For an APR1400 plant with two units, the usual shift crew composition includes five MCR operators (SS, STA, reactor operator, turbine operator, electrical operator), five local operators for each unit, and one operator for the common building shared by the two units.

The initial fire response team for an APR1400 unit is composed of at least five shift operators responsible for the initial detection and suppression of fires. The role of the team leader has to not be assigned to the shift supervisor (SS). Instead, a shift operator with either a reactor operator or SRO license needs to be designated as the leader.

At Darlington nuclear power plants in Canada, the composition of MCR crew at only a combined MCR covering four plant units together is composed of following operators:

— For plant operation, one shift manager; one control room shift supervisor; six authorized nuclear operators; two U0 control room operators;
— For fuel handling, one fuel handling control room supervisory nuclear operator; one fuel handling panel qualified operator.

Each position of plant operators in the MCR has Canada Nuclear Safety Commission certified role and the positions of fuel handling control room supervisory nuclear operator and fuel handling panel qualified operator are required anytime irradiated fuel is present in the FM head.

(b) ACP100 crew arrangement for main control room

In ACP100 as one of SMR models, the positions focused on the operation personnel who work in the MCR or the RSR. The staffing model for a MCR personnel configuration includes a shift supervisor, deputy shift supervisor, reactor operator A, reactor operator B, reactor operator C, and an isolation manager.

(c) BWRX-300 crew arrangement for main control room

The BWRX-300 is a 300 MW(e) water cooled, natural circulation SMR with passive safety systems. The minimum shift complement refers to the minimum number of station and emergency response workers who have to always be present to ensure the safe operation of the facility, respond to all credible events, and ensure adequate emergency response capability for the most resource intensive conditions.

The single unit plant minimum shift composition includes an at-the-controls operator in the MCR at all times, in all plant modes of operation. The operator has the training, qualification, competence, and expertise to take at-the-controls controls response in normal, abnormal, accident, and beyond design basis accident conditions. They initiate and direct site emergency actions until such time the emergency response team manager takes command and control.

The field staff includes fire escort, field and panel operations, and emergency mitigating equipment actions. They are trained in fire escort actions to fill the first emergency response (Primary function). They have the training, qualification, and competence to take simple at-the-controls actions in abnormal, accident, and beyond design basis accident conditions. They have the training, qualification, and competence to take manual field actions in abnormal, accident, and beyond design basis accident conditions. They are trained in maintenance activities to support emergency repairs. In the first seven days, they are credited with supporting fire/and or post-accident actions, including deploying emergency mitigating equipment.

The multi-unit plant minimum shift composition includes stand alone units within the same protected area with separate control rooms that are able to respond to common mode (loss of off-site power, seismic). The field staff includes fire escort, emergency mitigating equipment deployment, and emergency repairs. The operational staff (control room and field staff only) refers to the number of station workers expected to be needed to support the daily online plant programmes (surveillance maintenance, testing). This is estimated on preliminary information. On-site support staff (maintenance, chemistry, radiological protection, etc) are not included.

(d) KLT-40S crew arrangement for main control room

At the Akademik Lomonosov FNPP composed of two sets of KLT-40S units, Twelve personnel are working in an operation's shift such as a plant shift supervisor (PSS), two lead unit control engineers (LUCEs) (one per unit), a reactor department operator (RDO), a senior turbine department operator (STDO), two auxiliary turbine department operator (one per unit), an electrical engineer (EE), a plant equipment maintenance mechanic, an electrician for the plant high voltage equipment maintenance, an I&C engineer, and auxiliary operators of the RP equipment. Among the four operators working at a MCR overseeing both units simultaneously to conduct plant operations are a PSS, two LUCE (one per unit), and a STDO. The FNPP employs one hundred and sixteen (116) operating personnel in the aggregate, these staff are engaged in ensuring safe operation of the FNPP, ensuring nuclear and radiation safety; managing the operating modes of the plant, localizing and eliminating of accident consequences. The plant analyses the work, staffing, professional training of personnel, its certification, and obtaining permits from Russian regulatory body Rostechnadzor.

Rolls Royce SMR crew arrangement for main control room

Rolls Royce SMR has a single MCR to operate each SMR individually. It is expected that each MCR will be staffed by two or three personnel, the detailed assessment is continuing. The roles and training required for these operators are anticipated to be largely similar to conventional PWRs. The control rooms will include more software based controls and indications compared with operating plants. The systems operate with minimal operator intervention, including post-incident.

(e) SMART100 crew arrangement for main control room

A standard plant of the SMART100 has two units. Each unit is equipped with a MCR, and a shift in the MCR consists of one licensed SRO, one licensed reactor operator, two assistant operators and several local operators.

If operators need to evacuate the MCR due to such a fire and a loss of the MCR function, the operators transfer the control authority from the MCR to the RSR and move to the RSR as quickly as possible. In the RSR, the operators can bring and maintain the plant in safe shutdown condition in accordance with the appropriate procedure using the monitoring and control components mounted in the remote shutdown panel. An operation team in the RSR consists of one SRO, one reactor operator, two assistant operators, and several local operators.

(f) VOYGR^{TM} crew arrangement for main control room

The NuScale initially proposed a MCR minimum shift crew of six licensed operators in its recent design certification application. An integrated system validation was completed using the control room simulator and verified that the integrated system supports safe operation with performance based evaluations of hardware, software, and personnel. The NRC approved six licensed operator staff minimum. After reviewing the results of initial staffing plan validation efforts, NuScale conducted an additional study to evaluate a minimum shift crew of three licensed operators (two SROs and one reactor operator). Subsequent testing supported a topical report (TR-0420-69456-NP) with a revised staffing plan of a three-licensed-operator minimum staff with no STA [12]. This report was submitted in 2020 and approved by the US NRC and Advisory Committee on Reactor Safeguards in 2021.

In the VOYGR™ MCR, three licensed operators can safely operate up to twelve reactors in a single control room. The operating crew composition for the MCR includes a shift manager and control room supervisor (CRS), who may be combined or split depending on workload, a reactor operator 1, and an additional reactor operator. The license holders include one reactor operator and two SROs that are implemented in the technical specification 5.2.2 as one reactor operator and two SROs.

The STA position has been eliminated. The VOYGR™ 's revised control room staffing plan was audited by the NRC NRR branch in August and October 2020. The NRC issued a safety evaluation report endorsing the revised control room staffing plan (ADAMS Accession Number ML21160A131), stating that "the staff concludes there is reasonable assurance that the proposed minimum number of licensed operators is adequate to ensure safe operation of the plant. Therefore, subject to the conditions of applicability listed in Section 5.0 of this safety evaluation (SE), a NuScale facility licensee or COLA applicant may use the topical report as the technical basis for an exemption request from the staffing requirements". The NRC also issued SECY-11-0098 to inform NRC Commissioners that they endorsed the revised staffing plan for the VOYGR™ SMR and discussed how NRC processes have to be evolved to address the unique considerations of SMRs. The Advisory Committee on Reactor Safeguards performed an independent review of the NRC's evaluation of the VOYGR™ Control Room Staffing Topical Report in May 2021. The Advisory Committee on Reactor Safeguards issued a formal letter agreeing with the NRC staff endorsement [13].

The NuScale has demonstrated that its proposed minimum staffing complement can perform successfully in challenging operational scenarios without the use of a STA. The staff found that the STA position is not necessary for the safe operation of a VOYGR™ plant. The staff recognized that the STA position has been a valuable addition to the on-shift crew at operating reactors for over 40 years; however, using a performance based approach, they found that several elements collectively support the elimination of the STA for a VOYGR™ plant. These elements include the VOYGR™ control room human system interface design, the VOYGR™ plant design, the results of the revised staffing plan validation (RSPV) test, the availability of a second SRO on-shift, the licensed operator training programme, and that on-shift operators have time.

NuScale conducted a three-part mock NRC examination in October 2019 with nine NRC attendees from Region II and Headquarters. The examination consisted of simulator exams with three scenarios, a plant walkthrough exam with multiple pre-approved plant walkthrough questions, and a written exam.

NuScale presented its proposed format for a licensing exam at the NEI national operator licensing workshop in February 2020. They are also commercially committed to deploying NuScale technology and have an overview of their customer Utah Associated Municipal Power Systems.

For the HTR-PM composed of two NSSS modules and one turbine, four crew at the MCR conduct plant operation such as two reactor operators in charge of each NSSS independently, one conventional operator for the turbine and one supervisor for general coordination and management activities.

The assessment of verification activities includes observers who monitor and record human errors and behaviours such as communication with others. The operators are interviewed after the scenario to assess task difficulty, tension, and their thoughts on the scenario just completed. Plant parameters are also assessed to ensure they are within the threshold. Quantitative measurement includes workload and situation awareness. Scenarios with higher workloads include normal start-up and shutdown with many manual operations for reactor operators, multiple accidents for reactor operator/CO, and common cause failure for conventional operators. Scenarios with lower situation awareness include normal start-up/shutdown with many manual operations for reactor operator/Supervisor and multiple accidents for CO.

The analysis shows that there are inherent safety and passive safety features in place. There is no need for manual operations more than 48 hours after accidents. Operators only need to confirm the status. Necessary manual operation is only required if the protection system faults such as the manual reactor trip initialization and engineering safety features.

6.2. ROLES AND RESPONSIBILITIES OF MAIN CONTROL ROOM CREW

The MCR crew plays a vital role in the operation and safety of nuclear power plants. Their responsibilities are diverse and critical, requiring a high level of expertise and adaptability, especially with the ongoing transition including accident conditions. Some SMR plants adopt one MCR with minimum operating crew to perform the operation of muti units simultaneously with the support of passive safety systems, advanced automation and digital HMI such that the roles and responsibilities of MCR crew could be more critical than conventional nuclear power plants.

(a) APR1400 crew roles and responsibilities

In a APR1400 in Republic of Korea, the shift supervisor in the MCR has following assigned roles and responsibilities:

— Supervising other shift personnel in the conduct of operations and exercising command and control of shift activities per approved procedures and technical specifications;
— Being in charge of reactor safety, employee safety, public protection, direct power reduction operation, and reactor trip under uncertain conditions;
— Placing the plant in a safe condition using the auxiliary shutdown panel in case that the MCR is unavailable;
— Previewing and approving surveillance tests, maintenance activities, and temporary changes;
— Coordinate with other plant teams associated with chemistry, radiation protection, maintenance, I&C, and security activities;
— Supervising the plant operation support team activities required in the emergency plan, guidelines of health physics, and security plan;
— Classifying initial emergency alert levels and implement initial emergency actions required to mitigate an accident event until the emergency organization is set up;
— Performing other activities under the authority of plant director general and operation office director.

The shift technical adviser (STA) has duties as follows:

— Assisting in evaluating the operability of plant equipment per technical specifications;
— Providing oversight to ensure the crew is aware of the most recent procedures, plant design changes, and regulatory requirements;
— Maintaining an awareness of testing and maintenance activities that may impact the availability of the unit or the integrity of safety related equipment;
— Providing technical advice to the shift supervisor during normal, abnormal, accident, and transient conditions;
— Assessing plant parameters during and following an accident to ascertain whether core damage has occurred or appears imminent;
— Evaluating the effectiveness of procedures in terms of terminating or mitigating accidents and making recommendations when changes are needed;

— Investigating the causes of abnormal or unusual events, assessing any adverse effects, and providing recommendations on appropriate corrective actions;
— Providing an independent perspective of critical safety functions during transient and emergency conditions;
— Notifying transient and emergency conditions affecting the safe operation of the plant to the plant management and regulatory organization.

The reactor operator in the MCR of an APR1400 has following roles and responsibilities:

— Operating plant controls as directed by the shift supervisor per approved procedures and technical specifications;
— Monitoring plant operation parameters and responding to abnormal indications until corrected;
— Obtaining directions from the shift supervisor for the performance of evolutions;
— Reducing plant power to prevent plant shutdown and equipment damage;
— Manually shutting down the reactor if reactor protection system trip setpoints have been exceeded and a trip has not occurred;
— Manually initiating ESF if automatic ESF setpoints are exceeded;
— Ensuring that the unit is operating safely and take actions to mitigate events leading toward unsafe conditions (i.e., OLC violation, equipment damage);
— Directing local operator activities.

The turbine operator in the MCR is responsible for the operation of the secondary side of a nuclear power plant, including the feedwater system, steam system, turbine, condenser system, and instrument air systems. Finally, the electrical operator in the MCR is responsible for the operation of electrical systems including switchgear, motor control centres, uninterruptible power supplies, batteries, emergency diesel generators, and other electrical equipment.

During normal operation in four Darlington nuclear units covered by a MCR, each reactor unit will have an authorized nuclear operator at the controls. Additional authorized nuclear operators on shift will provide relief for lunches and breaks. Six authorized nuclear operators in the MCR have following roles and responsibilities:

— Continuously monitoring reactor unit on MCR panels;
— Operating or directing the operation of systems and equipment;
— Authorizing maintenance activities on the reactor units;
— Performing safety related system testing;
— Ensuring major automatic actions have occurred following safety system actuation;
— Ensuring critical safety parameters are monitored;
— Executing, or directing appropriate operating procedures necessary to respond to the event.

At least one of two unit 0 control room operators will be in attendance at the controls and conduct following assigned duties:

— Continuously monitoring unit 0 on MCR panels;
— Operating or directing the operation of common systems and equipment:
— Performing electrical switching for common services and reactor units;
— Authorizing maintenance activities on Unit 0 (common) systems;
— Performing safety related system testing;
— Ensure major automatic actions have occurred following safety system actuation;
— Ensuring critical safety parameters are monitored;
— Execute, or directing appropriate operating procedures necessary to respond to the event.

The fuel handling control room supervisory nuclear operator is also responsible for followings:

— Supervising MCR fuel handling personnel;
— Responding to alarms and/or abnormal conditions on fuel handling systems during fuelling activities;
— Recommending the appropriate course of action to control room shift supervisor in response to non-standard conditions;
— Contacting fuel handling technical when required.

The fuel handling panel qualified operator in the Darlington nuclear power performs followings:

— Continually monitoring fuel handling system panels anytime irradiated fuel is present in fuel handling machine head;
— Operating fuel handling systems in accordance with approved procedures to support refuelling activities;
— Notifying fuel handling control room supervisory nuclear operator of any abnormal conditions or alarms and perform corrective actions when directed.

A fuel handling control room supervisory nuclear operator and a fuel handling panel qualified operator will be present at the controls for any fuel handling System containing irradiated fuel.

The control room shift supervisor in the MCR has following duties:

— Supervising all control room activities;
— Authorizing increases in reactor power and maintenance activities on safety related systems;
— Authorizing the non-standard operation of fuelling machines when on a reactor or with irradiated fuel;
— Confirming the availability of station systems and equipment;
— Performing independent diagnosis and verifying that major automatic actions have occurred, critical safety parameter values are acceptable, and actions being taken are appropriate during plant transients.

The control room shift supervisor oversight is required for any reactivity changes and will also provide oversight for major equipment operation as work demands permit, but not mandatory.

The shift manager in Darlington nuclear power acts as a senior representative of station management on shift and conducts follows:

— Approving deviations from operating and maintenance procedures within the limits of authority specified in operating policies and principles;
— Attending and leading daily planning meetings and plant management activities;
— Acting as emergency response managers until formally relieved;
— Initiating consolidated nuclear emergency plans when required;
— Authorizing resetting of trips following safety system activation.

Authorized nuclear operators, Unit 0 control room operators, and fuel handling panel qualified operators will execute procedures and operate the controls independently. Second-party verification is required for reactivity device operation, high voltage switching, or non-standard operation on the fuel handling systems.

During transient operation in Darlington nuclear power plants, the lead authorized nuclear operator, control room operator and panel qualified operator will confirm the correct response of major automatic actions, diagnose, and execute the appropriate response procedure with control room supervisory nuclear operator concurrence for panel qualified operator. All control room staff will return to the MCR. Additional authorized nuclear operator will support transient units as panel operators taking direction from the lead. The lead will initiate monitoring of critical safety parameters. The control room shift supervisor will attend the incident unit, and independently confirm major automatic actions, critical safety parameters are monitored, and values are appropriate. The control room shift supervisor will also independently confirm diagnosis, and that the appropriate response procedure(s) are being completed. The shift manager maintains oversight of station activities and performs emergency operations centre activities as appropriate.

During multi-unit transients, the lead panel operators will confirm the correct response of major automatic actions, diagnose, and execute appropriate response procedures. All control room staff will return to the MCR. Additional authorized nuclear operators will support incident units based on severity. All available certified staff in the station will report to the MCR. The control room shift supervisor will take on the supervisory responsibility for the pair of adjacent units that are most impacted by the transient. Conversely, the shift manager will take on supervisory duties for the pair of adjacent units that are least impacted by the transient. The shift manager has to not only call in additional staff to support but also perform emergency operations centre activities.

(b) ACP100 crew roles and responsibilities

For the ACP100 in China, a shift crew of the MCR consists of six personnel who are a shift supervisor, a deputy shift supervisor, three reactor operators such as reactor operator A, B, and C, and an isolation manager. the SS carries out following duties during normal operation:

— Ensuring safe and stable unit operation as per laws, technical specifications, operation plan/procedures, and nuclear power plant production dispatch management systems;
— Being responsible for real time contact with grid dispatch department;
— Organizing relevant discipline personnel for analysis/diagnosis of sudden events and approving emergent handling of defects;
— Maintaining global situation with priority to safe operation conditions;
— Closely monitoring work affecting reactivity;
— Coordinating resources for handling transient state/accident as direct commander;
— Assuming duties in emergent response team as required and assisting emergency general command in relevant work.

The deputy shift supervisor in the MCR performs following roles and responsibilities;

— Assisting the SS in shift technical management to ensure safe and high efficiency nuclear power plant operation;
— Exercising SS duties upon temporary leave of the SS at the MCR;
— Maintaining an understanding of nuclear power plant's current status and parameters;
— Guiding operators to execute abnormal operation code and emergent operation code if needed by the nuclear power plant state.

The reactor operators are responsible for several key duties during normal operation as follows:

— Monitoring and carrying out system and equipment operations based on instructions from the SS and/or deputy shift supervisor;
— Guiding field crew operations when required, assessing issues identified during site inspections, and implementing appropriate solutions;
— Implementing measures to avoid surpassing safety limits in situations where safety limits are at risk.

The isolation manager conducts following duties:

— Conducting all operation isolation activities during shift;
— Leading the level 2 firefighting intervention team as the team leader to oversee and direct firefighting operations;
— Supervising field operations as the isolation manager.

During accidents, each MCR crew conducts followings:

— Shift supervisor confirms emergency level, broadcasting, reporting information to outside while assisting deputy shift supervisors to control accidents if necessary;
— STA who is not a part of the shift crew performs independent safety evaluation and oversight;
— Deputy shift supervisor implements emergent procedures general strategy, allocating/adjusting tasks to reactor operator A, B, and C, and supervises execution of operations;
— Reactor operator A executes mainline steps of the procedures;
— Reactor operator B conduct abnormal operation procedure required;
— Reactor operator C carries out operation summary (folded page) and continuous steps.

(c) BWRX-300 crew roles and responsibilities

The at-the-controls operator in a MCR of the BWRX-300 has a range of duties that they have to perform at all times in all plant modes of operation. They need to have the training, qualification, competence, and expertise to take at-the-controls response in normal, abnormal, accident, and beyond design basis accident conditions. They can initiate and direct site emergency actions until such time the emergency response team manager takes command and control (if shift manager and control room shift supervisor are not available). They are trained in first aid and fire escort actions to fill the first emergency response. They are also trained in maintenance activities to support emergency repairs. In addition, they are trained in effluent sampling and analysis as a backup to the online analyzer. Figure 12 presents a shift operation crew composition of the 4 units BWRX-300.

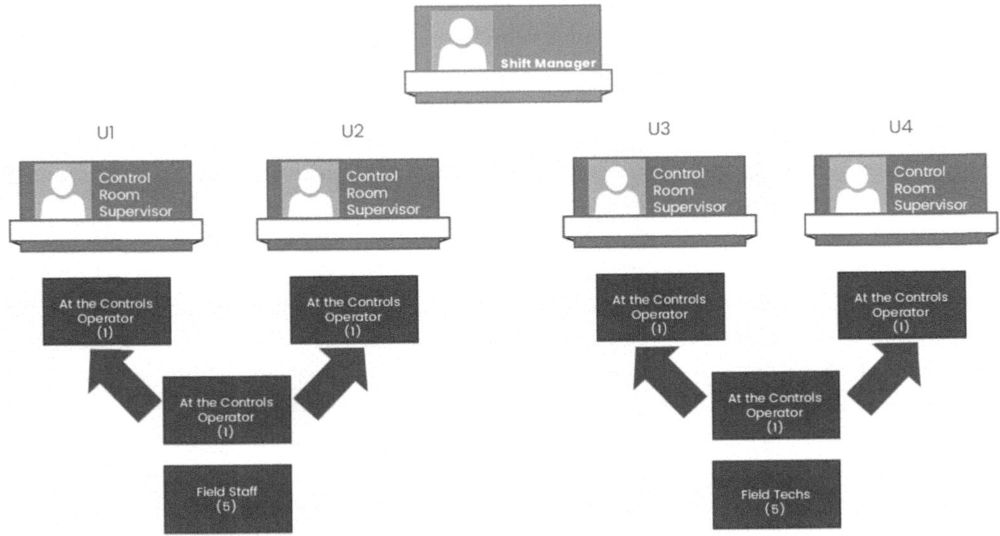

FIG. 12. A shift operation crew composition of the 4 units BWRX-300. (Source: GE Hitachi Nuclear Energy, United States, with permission)

Twelve personnel in a shift crew of the Akademik Lomonosov FNPP with both KLT-40S units conduct following roles and responsibilities:

— The PSS coordinates the safe and efficient operation of floating power unit equipment, ensuring adherence to the schedule for generating and supplying heat and electricity, as well as enforcing labour and production discipline among subordinate personnel;

— The LUCE is responsible for ensuring the safe and efficient operation of the reactor, steam turbine, main and auxiliary equipment, and systems within the designated service area in compliance with instructions, schedules, orders, and regulations;

— The RDO reports to the LUCE. The RDO informs the LUCE during shift turnover. Following the LUCE's directives, the RDO readies the reactor plant equipment for both units and the spent nuclear fuel facility from local positions for operation. They initiate operation, carry out maintenance from these positions, change the operation mode, shut it down, and monitor its status;

— The STDO, as directed by the LUCE, readies the steam turbine plant equipment for both units from local posts, starts operation, performs maintenance from these positions, changes the mode of operation, shuts down, and monitors the condition. They provide the LUCE with shift turnover updates, results from periodic equipment and premises inspections, and log entries in the operational documentation;

— The auxiliary turbine department operator, following instructions from the RDO and STDO, readies turbine equipment for both units from local posts, initiates operation, conducts maintenance from these positions, adjusts the operation mode, shuts down, and monitors the condition. They update the RDO and STDO during shift turnover, provide feedback on equipment and premises inspections, and document entries in the operational reports;

— The Electrical Engineer (EE) follows industry regulatory documents, technical specifications, and operating procedures while overseeing EE production activities. Operating from the automated workstation under the PSS guidance, the EE manages the technological process within the grid. They ensure adherence to the power supply schedule and execute switching operations in grids after coordinating with the PSS, RDO, and STDO. Additionally, the EE is responsible for supervising the safe and reliable operation of the grid in both normal and emergency modes;

— The power plant equipment maintenance mechanic follows instructions from the PSS to ready equipment for operation, start operation, conduct maintenance from local positions. They provide the PSS with feedback on periodic equipment and premises inspections during shift turnover and log entries in the operational documentation as needed. In addition, the mechanic promptly reports any abnormal equipment operation modes or emergency situations to the PSS and enacts corrective measures in alignment with operating procedures;

— The Electrician responsible for high voltage equipment maintenance reports to the electrical engineer (EE) during shift turnover, providing updates on equipment and premises inspections. Following the PSS instructions, the electrician readies high voltage equipment for operation, puts it into operation, conducts maintenance from local positions. In case of parameter deviations in electrical equipment or emergencies, the electrician immediately notifies the EE, takes corrective actions as per operating procedures, and contributes to accident reviews related to their equipment;

— I&C engineer ensures operation, maintenance and repair of reactor control and protection systems, automation and I&C, information and computer systems (ICS) in accordance with the rules, instructions, orders, based on the organization of relevant work, controls the technical condition and safe operation of equipment, participates in the investigation of the causes of its failure, keeps records, analyses equipment failures, comments on its operation, and summarises operating experience;

— The auxiliary operator of the reactor protection equipment reports to the RDO and STDO during shift turnover. Acting on directives from the RDO and STDO, the auxiliary operator readies the reactor protection equipment and spent nuclear fuel for both units from local posts.

(e) SMART100 crew roles and responsibilities

In the SMART100, the MCR is designed to provide operational flexibility to accommodate a wide range of MCR staffing requirements. A shift in the MCR consists of one licensed SRO, one licensed reactor operator, two assistant operators and several local operators. The SRO, reactor operator and two assistant operators are always in the MCR. Each operator's role in the MCR is as follows:

— The SRO supervises operators and coordinates the works among operators and cooperates with the operators. Also, the SRO approves operations of the systems that significantly influence the operation of the SMART100 in every operation mode. The SRO is in charge of the whole work of a shift crew;
— The reactor operator is responsible for the operation of primary and secondary systems during normal operation. The reactor operator implements the instructions of the SRO and monitors and controls the operation parameters for all operation modes and reports the results to the SRO. The reactor operator operates the reactor and related systems based on the support of assistant operators during abnormal or emergency operation;
— Two assistant operator support the entire administrative works during normal operation. One assistant operator 1 is in charge of turbine operation and the other assistant operator 2 is in charge of the operation of the whole systems and supports the operation of reactor and turbine. Also, the assistant operators help the reactor operator for accurate awareness of the state of the plant through the analyses of various kinds of transient. In addition, the assistant operators support the reactor operator according to the instructions of the SRO for safety related actions as a shift technical assistant (STA);
— Local operators access the local control stations from time to time through the communication with the MCR operators and operate the local control panels. The local operators perform their tasks at the local control panels during abnormal and emergency operation and support the MCR operators through continuous communication with them.

When the MCR is unavailable, the operators in the MCR can move to the RSR and shut down the reactor using the remote shutdown panel in the RSR. Each operator's role in the RSR is as follows:

— The SRO gives instruction on corrective actions to operators when abandonment conditions occur and checks whether the corrective actions are completed. Furthermore, the SRO moves to RSR and performs his or her duties. The SRO monitors all plant overall conditions, gives instructions, and receives the report after safe shutdown actions are taken. The SRO monitors safe shutdown conditions in the SMART100 until all tasks of reactor operator and assistant operators are completed in the RSR;
— The reactor operator trips the reactor according to the instruction from the SRO and checks the reactor shutdown status upon abandonment in the MCR. Once shutdown is completed, the reactor operator moves to the RSR quickly. In the RSR, the reactor operator confirms the transfer of the control authority on the remote shutdown panel. Then, the reactor operator manipulates the monitoring and control components installed on the remote shutdown panel and performs safe shutdown. Until the tasks are completed in the RSR, the reactor operator monitors the safe shutdown conditions of the plant, follows the instructions from the SRO, and reports the occurrence of any abnormal condition;
— The assistant operator 1 receives instructions from the SRO and runs the remote shutdown breaker in the electric equipment room when reactor shutdown is not possible in the MCR. The assistant operator 2 receives instructions from the SRO and moves to the I&C equipment rooms to transfer the control authority from the MCR to the RSR and then move to the RSR quickly. The assistant operators assist the SRO with regard to safety related safe shutdown actions. The assistant operators perform the required safe shutdown actions based on the instruction from the SRO and reports the results;
— Local operators receive instructions from the SRO and shutdown the turbine using the manual shutdown lever on the front side of the turbine at the site when turbine shutdown cannot be done in the MCR. Then, they perform the required safe shutdown actions according to the instruction from the SRO and report the results.

The composition of MCR staff in the VOYGR™ has been optimized by leveraging passive safety systems of VOYGR™, fail-safe design features, high levels of automation, and minimal important human actions. The staffing level was selected to be safe and reliable when operating a 12 unit plant. Roles and responsibilities for a VOYGR™ MCR staff are as follows:

— The shift manager is in charge of overall shift operations. The shift manager is knowledgeable in all plant disciplines and ensures that the duties of the chemistry, health physics, instrumentation, and other maintenance support services are performed as needed for safe plant operation. The shift manager is the senior licensed operator assigned to the MCR team and acts as the senior manager on site when the plant manager and operations manager are not available. The shift manager is the initial person responsible to implement the emergency plan. The emergency plan responsibilities have to be maintained until properly relieved in accordance with the station emergency plan requirements. The shift manager acts as the conduit between station management and the on-shift plant staff. The shift manager holds an NRC SRO license and meets the requirements of ANSI/ANS-3.1-2014 for the shift manager position. This position is combined with the CRS when there are only three licensed operators on site;

— The Control Room Supervisor (CRS) command and control of the MCR, resides with the control room supervisor. The CRS is responsible for all units and directs and oversees the activities of the licensed and non-licensed operators. The CRS is also responsible for authorizing activities that impact plant operations. The CRS is responsible for ensuring the appropriate staff is available in the MCR to manage the workload. The CRS has the authority to shut down units that are presenting an undue burden to the crew as a tool to manage workload. The CRS always has the authority to direct resources or activities associated with operation of the plant. The CRS maintains and enforces the standards of conduct in the MCR. The CRS holds an NRC SRO license;

— The reactor operator 1 is primary responsible for monitoring all 12 units and the balance of plant systems and responsible for prioritization of notifications and all normal reactivity changes including power changes and power maintenance. This operator maintains situational awareness of all units under the authority of the control room and performs routine tasks to maintain steady state operation. The reactor operator 1 provides the initial response to any unexpected off-normal conditions, for example fire alarms and can perform simple tasks that do not significantly detract from the overall monitoring function. The reactor operator 1 generally responds to all notifications and determines the appropriate level of action. The reactor operator 1 may peer-check manipulations performed by the other reactor operators but has to not be involved in activities that would require longer term focused attention. reactor operator 1 will hold a reactor operator or SRO license;

— An additional reactor operator has to be assigned to the control room to perform the shift surveillances and support required testing and maintenance for all modules and the common systems. This includes supporting maintenance activities that require control room manipulations. Reactor operator 1 will generally delegate an unexpected alarm response to the additional licensed operator. The additional reactor operator will hold a reactor operator or SRO license;

— Non-licensed operators are not included in the MCR staff but are responsible for operation outside of the MCR including system line-ups, tagging, investigation, and fire response as directed by the MCR staff.

6.3. HUMAN MACHINE INTERFACE

The HMI in a nuclear power plant is a critical component that allows operators to monitor and control the state of the plant. The evolution of HMI in nuclear power plants has seen a transition from analogue to digital systems that has significantly changed the way operators interact with the systems. HMIs in nuclear power plants has evolved significantly over the years, with digital technology playing a key role in enhancing the efficiency and safety of nuclear power plant operations. However, the design of the HMI needs to carefully consider the potential challenges to ensure safe operation.

(a) APR1400 human machine interface

The HMI in APR1400 is designed to make use of advanced digital technologies that have been proven through industrial operating experience and extensive testing and development programs. The HMI designs encompass the design of the MCR and the plant I&C systems. As presented in Fig. 13, the MCR layout of APR1400 consists of functions distinct from those of analogue control rooms such as the information display workstation, the overview display panel, and the alarm summary display panel. The information display workstation is used to display various types of information, including process variables, alarms, and status information. The overview display panel is used to display an overview of the plant status, including the status of major systems, equipment, and components. The alarm summary display panel displays a summary of all active alarms in the plant.

Additionally, a large panel display in MCR is designed to provide an overview or high level summary of the plant status and safety related information to the operating crew. It consists of five major sections including a fixed mimic section with the reactor operator and turbine operator sections and process & instrumentation diagram layout, the large display panel variable display section that offers a useful facility for the presentation of process information on a less permanent basis, and overview information required for plant operations change based on plant operating conditions and the needs of the operating crew, and an operator console in MCR.

FIG. 13. Main Control Room panel layout of the APR1400. (Source: KHNP, Republic of Korea, with permission).

The safety console in the MCR has critical functions and features. It has a minimum inventory of fixed position controls and preferred/credited success path components in a major flow path for each critical safety function. It also has components required to perform safe shutdown and critical tasks identified by the probabilistic safety assessment and human reliability analysis (HRA). The qualified displays and alarms support Class 1E controls. The safety console also has a diverse manual ESF actuation switch & diverse indication system flat penal display performing hot shutdown and critical safety functions. The ESF component control system soft control module and conform switch are also present in the safety console along with turbine backup flat penal displays, Plant protection system, core protection calculator modules, and mini large display panels.

(b) KLT-40S human machine interface

In the KLT-40S, HMI management arranged at the MCR with four crew members on it, the Akademik Lomonosov FNPP has a single MCR for both NSSS units shown at Fig. 14. below.

FIG. 14. Configuration of main control room in the Akademik Lomonosov. (Source: Afrikantov OKBM, Russian Federation, with permission).

In Fig. 14, the rear panel, not visible, is utilized by the PSS. The front panels are segregated into three zones: two panels on the left and right sides are designated for the LUCEs, while the panel in the centre is used by the STDO. The automated workstation of each LUCE features a mimic panel alongside four screens. The automated workstation of the STDO is equipped with four screens and two recorders. Control keys are positioned on the horizontal surfaces of the panels, while additional I&Cs are placed on the vertical boards of the panels to facilitate power unit control. If the MCR becomes inaccessible, the MCR crew relocate to the supplementary control room where all essential I&C are available to safely initiate the reactor shutdown process.

In the SMART100 design, the MCR is provided to enable the SMART100 plant to be operated safely in all operation modes and to bring to a safe state after accidents. Such design basis events are considered in the design of the MCR. As presented in Fig. 15 below, the MCR has the following workstations and the large display panel such as following:

— MMCW consists of an integrated workstation including the individual workstation for the SRO, reactor operator and assistant operator;
— Auxiliary monitoring and control workstation provides functions to monitor and control the heating, ventilating and air conditioning system, power supply and electrical system and radiation monitoring system;
— SMCW is equipped with displays and controls for monitoring reactor safe shutdown and for responding to accidents. The SMCW provides operational information and controls to enable continuous operation during functional loss of the MMCW. The SMCW is designed to maintain minimum inventory for safe shutdown and emergency operation in case of total failure of the MMCW. The SMCW is allocated near the reactor operator console of the MMCW in order to guarantee quick accessibility to the SMCW during an emergency operation;
— Large display panel is installed at the front of the MMCW so that all operators residing in the MCR during all operation modes including normal and abnormal operation can access the plant overview information easily while not only seating at the MMCW or the auxiliary monitoring and control workstation but also standing anywhere.

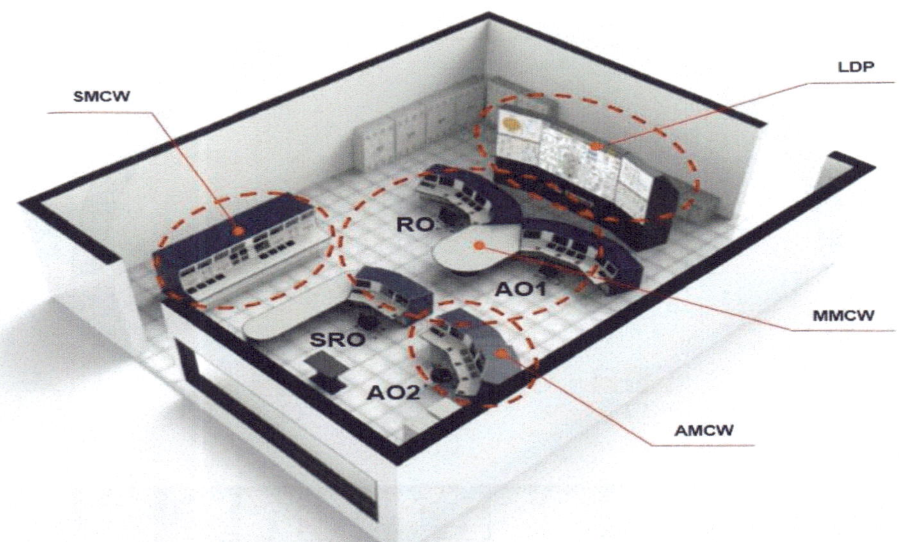

FIG. 15. Configuration of main control room in the SMART100. (Source: KAERI, Republic of Korea, with permission).

When the MCR becomes unavailable, the operators move to the RSR that is physically separated and electrically independent from the MCR. The operators can shut down the reactor using the remote shutdown panel in the RSR. The remote shutdown panel consists of the safety grade soft controllers, the information displays including non-safety grade soft control functions, the indication display, the alarm display, the alarm sound component, the manual trip and actuation switches, and the control transfer switch. The environmental design of the RSR is the same as the MCR design since the environmental design factors of the RSR could have a large influence in the performance of operators.

(c) VOYGR^{TM} human machine interface

The MCR of VOYGR™ plant as presented in Fig. 16 below contains the following equipment and features:

— A bank of video display units (VDUs) configured with safety display and indication system of human system interface;
— Sit-down workstations for multiple licensed operators, each able to access human system interfaces for all units and common systems;
— A dedicated stand-up control panel for each unit;
— A dedicated stand-up control station for shared or common systems;
— A dedicated manual control for safety system actuation, component repositioning, and overriding of specific safety signals in severe accident conditions.

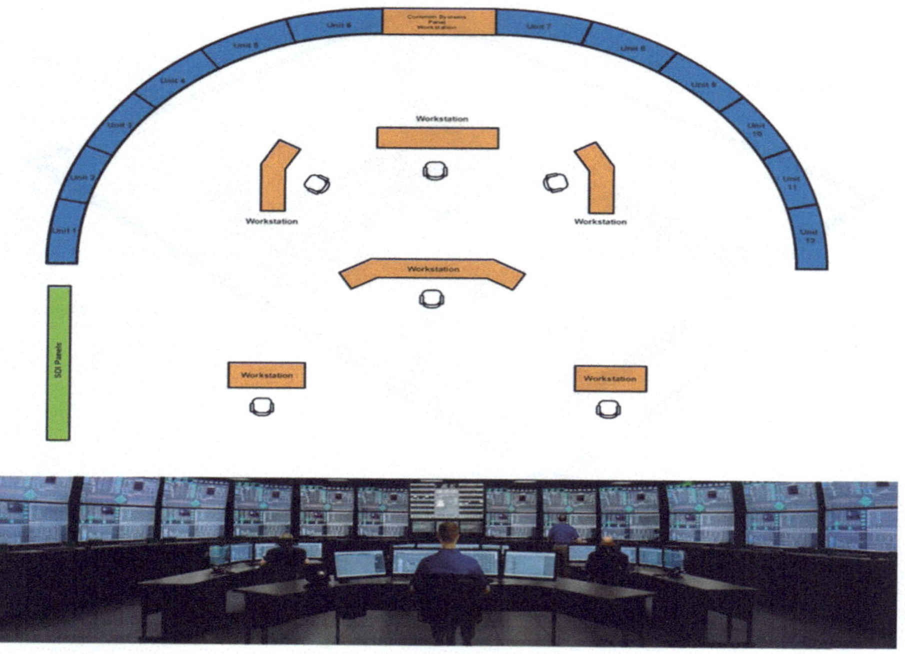

FIG. 16. The MCR layout and concept of the VOYGR™-12. (Source: NuScale Power Inc., United States, with permission).

The workstations and the common control system panel of VOYGR™ have following assigned features as follows:

(i) Sit-down workstations

One operator attends a sit-down workstation that comprises four VDUs. The human system interfaces displayed on the VDUs are navigable. The VDUs contain the alarms, controls, indications, and procedures necessary to monitor and manage any unit chosen by the operator during normal, abnormal, emergency, shutdown, and refuelling operations. The MCR operators and supervisors' interface with the plant at their designated workstations using human system interface software located on the plant control system and module control system networks. Due to high levels of automation and passive safety functions, multiple units may be controlled by a single operator at any workstation simultaneously. Furthermore, common or shared plant systems can be fully monitored and managed from each workstation. The capability of the human system interface and the supporting plant control system and module control system network architecture structure allows the operator workstations to support oversight and control activities.

Figure 17 below is an example of a sit-down workstation.

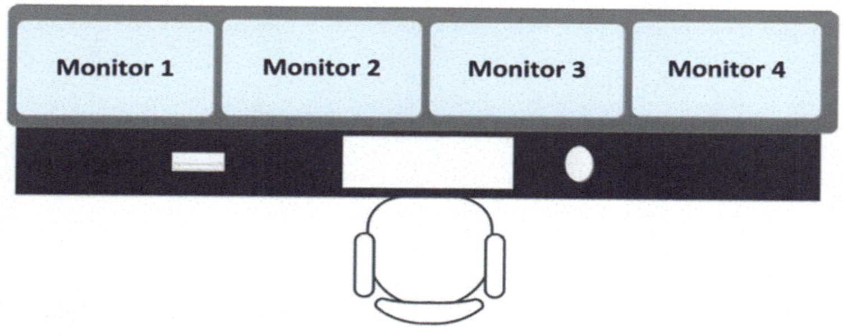

FIG. 17. A sit-down workstation of VOYGR™. (Source: NuScale Power Inc., United States, with permission).

(ii) Standup workstation

During most operating conditions, the uppermost larger unit workstation VDU (i.e., unit overview display) provides an overview display for that unit so that other MCR personnel can quickly determine status. The synchronized data control capabilities of the unit control system allow an operator to perform more dedicated specific unit activities on each of the stand-up unit workstations. The stand-up unit workstations are specific to each unit and include hard wired safety system actuation switches and non-safety enable switches.

Figure 18 below is an example of a standup workstation.

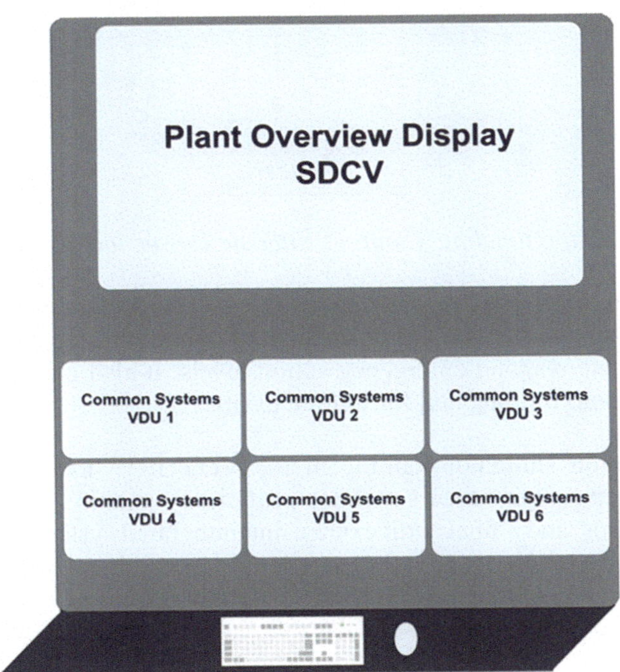

FIG. 18. A standup workstation of VOYGR™. (Source: NuScale Power Inc., United States, with permission).

(iii) Common control system panel

The front-facing central overview display in the MCR presents a VOYGR™ plant wide overview with specifically selected plantwide monitoring items visible to any operator in the MCR. Like the stand-up unit workstations, the function of the VDUs may be accomplished by other means.

The human system interface layout in the MCR is specifically designed to support minimum, nominal, and enhanced staffing during all operating plant modes. Shared system VDUs and unit/plant overview VDUs are located so they can be observed from multiple locations in the MCR. Unit workstations are spaced so that side-by-side operation at adjoining units allows sufficient room to maneuverer.

NuScale's human factors engineering timeline includes a 12-unit MCR simulator commissioned in 2012, a staffing plan validation in 2016, integrated system validation in 2018, and revised staffing plan validation in 2019. As a result of the integrated system validation training experience, NuScale is confident that a commercial initial license class for a control room operator can be completed in nine months or less as shown in the Fig. 19 human factors engineering timeline below.

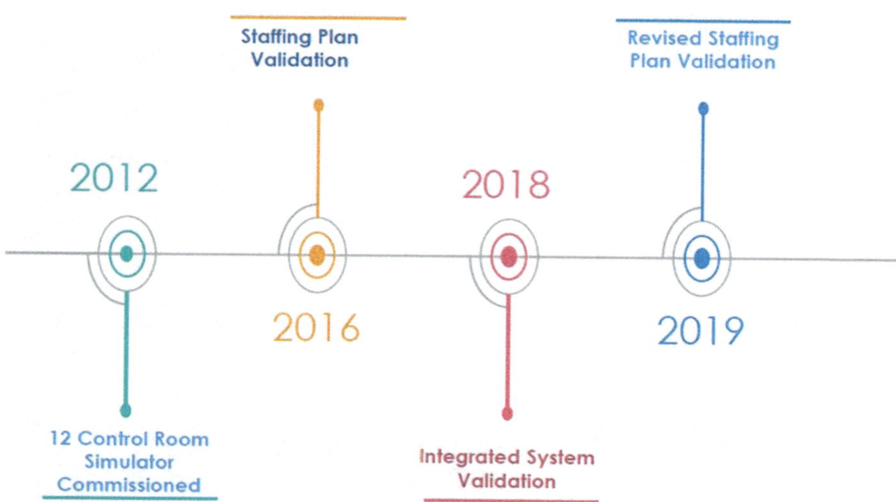

FIG. 19. Human factors engineering timeline. (Source: NuScale Power Inc., United States, with permission).

The integrated system validation classroom training in the VOYGR™ includes 56 (fifty-six) classroom lectures over nine weeks including systems training, conduct of operations, technical specifications, abnormal and emergency operating procedures, and emergency action levels. It also includes eleven simulator review sessions to complement classroom training and for written exams.

The integrated system validation simulator training in the VOYGR™ includes 36 (thirty-six) simulator sessions for each crew over ten weeks, three simulator proficiency exams, one week of integrated system validation style practice scenarios, and a final audit exam using integrated system validation protocols. Insights from simulator evaluations on VOYGR™ show that simple design and human factors are integrated into the human system interface that reduces human errors (no miss-operated equipment in all the testing that has been completed to date), makes operation intuitive, enhances safety, reduces training time, and simplifies examination.

(d) HTR-PM human machine interface

The safety console of the HTR-PM in China has HMIs for different modules that are separated. The VDU displays are divided into three categories according to the function analysis: displays for reactor #1, displays for reactor #2, and displays for steam turbine.

The VDU displays in the HTR-PM have special designs such as customized menus in the header that provide access to all displays, system tools, access to alarm list displays of different types, and workstation indication. The access to alarm list is shown in the header of each display. When one menu is clicked, the display list of the corresponding type is shown as a popup menu. For the operator at the workstation of reactor #1, the display lists on the popup menus are all about reactor #1. The menus of reactor #2 and conventional island (CI) are similar. The font size is big to identify the number of modules and avoid confusion between the two modules. The title and name of each display are clearly shown in the bigger size. The naming rule of display specifies that the first number of the display name is used to identify the module, such as '1' for reactor #1 and '2' for reactor #2.

The approach to cope with specific human factors engineering issues in the HTR-PM include an authority for control position. There are four control positions in the control room. The control authority is designed to correspond to the control position. All the displays in the system can be viewed through any monitor in the control room. However, the controls of equipment are designed to be authorized by assigned operators and the specific workstations. The control logic is such that if the system is used by one reactor, the other reactor is unable to use the system. Human factors engineering consideration is also taken into account where the control and switchover need to be confirmed by all operators and needs to be noticed in training and procedure development.

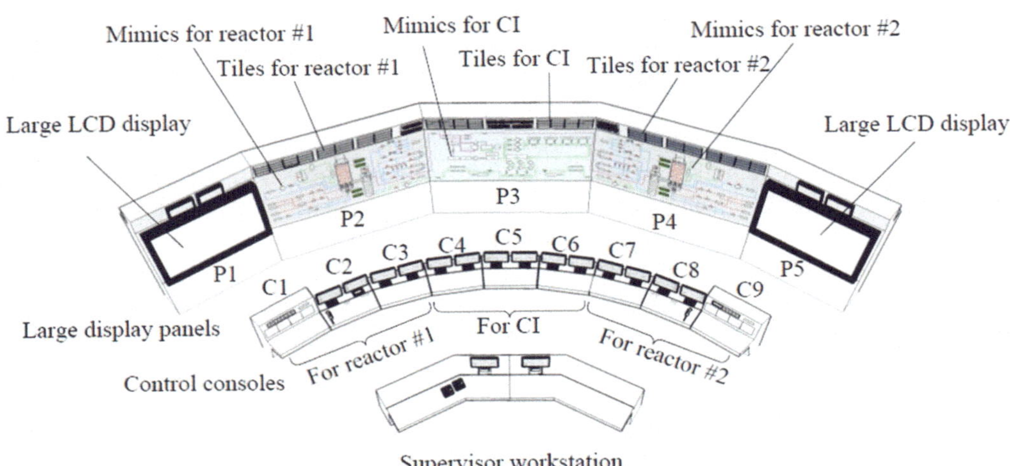

FIG. 20. layout of human machine system in MCR of the HTR-PM. (Source: Huaneng Group, China, with permission).

As it can be observed in Fig. 20, the MCR display panels of HTR-PM are in the front of the control room and include large screen displays, mimic displays, and alarm tile displays. They provide an overview or system level summary of the plant status and support crew coordination and quick awareness of the plant changes and personnel communication and collaboration for multi-person control tasks.

The MCR mimic displays in HTR-PM provide an overview of the plant and are spatially dedicated continually visible displays that allow for rapid detection and enhanced pattern recognition. The position of the mimic display is fixed and cannot be reconfigured or modified. The operator can have access to information easily without interface management. The position of each device on the mimics is visible at a glance. As a diversity of the screen display, the control signals of the indicators on the mimics are generated from local equipment or control cabinets. In this regard, the mimic displays maintain basic monitoring of the entire plant in the condition where all non-safety screen displays fail due to common cause failure of network or computer software, etc.

Verification and validation include a verification platform of 1:1 and verification and validation activities that follow the guidelines and standards in China. Different verification and validation activities are performed at different stages and are approved by the nuclear safety review in China.

7. SUMMARY AND FURTHER CONSIDERATIONS

Currently, over 70 SMR designs and technologies are under different phases of development and deployment. Countries not only expanding but also embarking in nuclear industry, interested in synergy among nuclear and distributed energy resources including renewables, expect high potential of SMRs contributing to achieving net zero emission by 2050.

The IAEA and other international organizations have issued publications on SMRs in the areas of technology, fuel cycle, economics, safety, safeguards, security and infrastructure development. However, the resources on SMR operations are still limited. Through various IAEA meetings, Member States highlighted their need for up-to-date technical information and experience in the SMR operations. Such information is crucial during the current phase of SMR development and deployment.

Through a series of meetings on the approach and preparation for operation of SMRs, key aspects of plant operations and control and maneuvering, fuel integrity and operating personnel management as well as HMI of MCR in SMRs were discussed in comparison with those of conventional large nuclear power plants. Features of existing large nuclear power plants such as a combined MCR for Darlington nuclear power plants in Canada and load following operation with combination of EDF nuclear fleet in France and digitized HMI in the APR1400 in Republic of Korea are used as useful references to understand the operating characteristics of SMRs. Therefore, comparing those features between large nuclear power plants and SMRs are practical means to understand operating characteristics of SMRs.

SMRs represent a promising evolution in nuclear technology. Many of the designs for near-term deployment incorporate design simplification with scaled-down size. Water cooled SMRs benefit from the operating PWR and BWR fleets, incorporating decades of operational experience and safety performance derived from the broader nuclear field. Non-electric applications of SMRs, such as hydrogen production, desalination, and industrial heat supply, highlight their potential to contribute significantly to a low carbon economy, offering solutions beyond traditional power generation. The modularity and flexibility of SMRs not only aid in reducing construction times and costs but also enables these reactors to support a wide range of energy strategies, adapting to fluctuating power demands and integration with renewable energy sources.

Nuclear power plants with SMRs operate under typical operational modes of existing nuclear power plants such as power operation, startup, hot shutdown, cold shutdown, and refueling. These operational modes provide a framework on how SMRs would respond under various conditions and during different phases of operation. The OLCs are critical in defining the parameters in which nuclear plants have to operate to ensure safety. These limits include specifications for equipment operability and specify the necessary conditions to ultimately preclude radioactivity release.

The operational testing phases, such as cold hydrostatic tests and power ascension tests, play a vital role in validating reactor design, construction, and readiness level. These tests are crucial in assessing the integrity of systems under operational conditions before commercial operation of the reactors. Each reactor model includes specific operational procedures tailored to its design and technological features, ensuring that each step from startup to full power operation is conducted within safe and controlled conditions, as predetermined by rigorous testing and safety evaluations.

Regarding SMR control and operations, SMRs incorporate NSSS features crucial control systems such as the plant control system that ensures operational safety and efficiency through automation and integration in power level controls. SMR systems also incorporate specialized controls for NSSS. Some SMRs leverage natural circulation within the reactor for simpler, safer control, automatically regulating reactor pressure to drive turbine generators using the steam flow directly from the RPV. Other SMR designs focus on inherent safety using negative reactivity coefficients and automated control systems for reactor power and coolant adjustments, enhancing adaptability and operational flexibility to accommodate fluctuating grid demands and support renewable energy integration. Regardless the various designs and technologies, SMRs are equipped with sophisticated control systems that allow for safe, efficient, and flexible operation, adapting to modern energy demands and supporting the transition to low carbon energy systems.

Water cooled SMRs leverage proven LWR fuel designs with enrichments typically under 5%, while marine based SMRs may employ a variety of fuel types with varying enrichment levels, presenting a broad array of design and operation parameters.

On advanced fuel cycles, exploring high assay low enriched uranium for longer fuel cycles and higher efficiency have been conducted since early 2010s. However, there are challenges in handling high burnup fuels and safety concerns such as fuel fragmentation during accidents. Issues such as thermal hydraulic impacts on fuel integrity, potential for control rod ejection, and economic efficiency due to lower discharge burnup are critical considerations for ensuring the operational viability and safety of SMRs. The complex design of some SMR fuel assemblies necessitates meticulous verification processes and management strategies to handle rapid power changes and large axial power offsets that could impact fuel performance and overall reactor safety.

SMRs are generally designed to support flexible operation crucial for integrating with variable renewable energy sources. Understanding and managing the impacts of such operational flexibility on fuel integrity is essential for maintaining safe reactor operations. From water cooled to other than water cool reactors, each SMR type adopts unique fuel strategies to optimize safety, efficiency, and adaptability to reflect the innovative approaches being explored in the nuclear energy industry.

The composition of the MCR crew varies between large nuclear power plants and SMRs, with large nuclear power plants typically employing more personnel due to their scale and complexity. The SMRs, on the other hand, benefit from increased automation and reduced number of operators, enhancing efficiency. In SMRs, roles are consolidated due to the reduced scale and enhanced automation. Duties are distributed among fewer operators who manage multiple aspects, highlighting the versatile capabilities required in smaller, more technologically advanced control rooms. Each plant type equips its MCR crew with distinct protocols for handling emergencies, including fire response and accident mitigation, ensuring rapid and effective response capabilities.

Advanced HMI systems are integrated within MCRs across different plant models, assisting operators with effective monitoring and management of plant operations, enhancing safety and operational efficiency. Designers of SMRs pursue strategic staffing models that undertake operational demands and potential emergencies, ensuring that the MCR is always equipped with adequately trained personnel to handle routine and extraordinary situations. The design of MCRs is evolving to accommodate digital human and machine interface features that aid operators in maintaining plant safety and efficiency, adapting to modern operational standards and technological advancements.

Dissemination of information on operation approach and experience from the existing operating SMRs and near term deployable designs can help identify potential design deficiencies of first-of-a-kind SMRs at early stage of design development.

REFERENCES

[1]	INTERNATIONAL ATOMIC ENERGY AGENCY, Advances in Small Modular Reactor Technology Developments 2022 Edition, IAEA, Vienna (2022).

[2]	INTERNATIONAL ATOMIC ENERGY AGENCY, Non-baseload Operation in Nuclear Power Plants: Load Following and Frequency Control Modes of Flexible Operation, IAEA Nuclear Energy Series No. NP-T-3.23, IAEA, Vienna (2018).

[3]	INTERNATIONAL ATOMIC ENERGY AGENCY, IAEA Nuclear Safety and Security Glossary 2022 (Interim) Edition.

[4]	US NRC Regulations 10 CFR 50.36 Technical specifications (2021).

[5]	INTERNATIONAL ATOMIC ENERGY AGENCY, Operational Limits and Conditions and Operating Procedures for Nuclear Power Plants, IAEA Safety Standards Series No. SSG-70 Vienna (2022).

[6]	US NRC NUREG-1434, Standard Technical Specifications – General Electric BWR/6 Plants, Revision 5 (2021).

[7]	INTERNATIONAL ATOMIC ENERGY AGENCY, Safety of Nuclear Power Plants: Design, IAEA Safety Standards Series No. SSR-2/1 (Rev. 1), IAEA, Vienna (2016).

[8]	US NRC regulation guide 1.97, Revision 5, Criteria for Accident Monitoring Information for Nuclear Power Plants (2019)

[9]	US NRC regulation guide 1.68, Revision 3, Initial Test Programme For Water-cooled Nuclear Power Plants (2013)

[10]	EPRI Advanced Nuclear Technology: Advanced Light Water Reactor Utility Requirements Document (Revision 13)

[11]	Methodology for the Systematic Assessment of the Regulatory Competence Needs (SARCoN) for Regulatory Bodies of Radiation Facilities and Activities, IAEA-TECDOC-1860, IAEA, Vienna (2019).

[12]	Topical Report NuScale Power, LLC Submittal of 'NuScale Control Room Staffing Plan', TR-0420-69456, Revision 1 (2020)

[13]	US NRC SECY-11-0098 Operator Staffing For Small or Multi-Module Nuclear Power Plant Facilities (2011)

ABBREVIATIONS

APR1400	Advanced Power Reactor 1400
CAREM	Central ARgentina de Elementos Modulares
COLA	COmbined License Application
CVCS	Chemical Volume Control System
EDF	Electricité de France
FNPP	Floating Nuclear Power Plant
HFT	Hot Functional Test
HMI	Human Machine Interface
HTR-PM	High Temperature Reactor Pebble-bed Module
I&C	Instrumentation and Control
KAERI	Korea Atomic Energy Research Institute
LUCE	Lead Unit Control Engineer
LWR	Light Water Reactor
MCR	Main Control Room
MMCW	Main Monitoring and Control Workstation
Nnom	Rated Electric Power Capacity
NPM	NuScale Power Module
NSSS	Nuclear Steam Supply System
OLC	Operational Limits and Condition
PCI	Pellet Cladding Interaction
PSS	Plant Shift Supervisor
RCP	Reactor Coolant Pump
RCS	Reactor Coolant System
RDO	Reactor Department Operator
RPV	Reactor Pressure Vessel
RSR	Remote Shutdown Room
SCC	Stress Corrosion Cracking
SDA	Standard Design Approval
SMART	System-Integrated Modular Advanced Reactor
SMCW	Safe Shutdown Monitoring and Control Workstation
SMR	Small Modular Reactor
SRO	Senior Reactor Operator
STA	Shift Technical Advisor
STDO	Senior Turbine Department Operator
TRISO	Tristructural-Isotropic
VDU	Video Display Unit

CONTRIBUTORS TO DRAFTING AND REVIEW

Ceyhan, M.	International Atomic Energy Agency
D'Souza, N	Ontario Power Generation (OPG), Canada
Despretz, O.	International Atomic Energy Agency
Feutry, FS.	Electricité de France (EDF), France
Flamand, R.	NuScale Power, United States of America
Gooh, Y.S.	Korea Hydro and Nuclear Power Company (KHNP), Republic of Korea
Han, J.	China Huaneng Nuclear Group (CHNG), China
Hinds, D.	GE Hitachi Nuclear Energy, United States of America
Hunnewell, S.W.	Tennessee Valley Authority (TVA), United States of America
Janin, D.	Bain & Company, Denmark
Jung, J.H.	International Atomic Energy Agency
Kim, S.	Korea Atomic Energy Research Institute (KAERI), Republic of Korea
Kim, W.B.	International Atomic Energy Agency
Lyu, S.	Sanmen Nuclear Power Plant, China National Nuclear Power (CNNP), China
Marquez, M.	NuScale Power, United States of America
Renev. A.	Rosatom Energy Projects JSC (REP), Rosatom, Russian Federation
Schlamp, M.	Comisión Nacional de Energía Atómica (CNEA), Argentina
Subki, H.	International Atomic Energy Agency
Tovar, T.	NuScale Power, United States of America

Technical Meeting
Vienna, Austria: 22–25 August 2023

CONTACT IAEA PUBLISHING

Feedback on IAEA publications may be given via the on-line form available at:
www.iaea.org/publications/feedback

This form may also be used to report safety issues or environmental queries concerning IAEA publications.

Alternatively, contact IAEA Publishing:

Publishing Section
International Atomic Energy Agency
Vienna International Centre, PO Box 100, 1400 Vienna, Austria
Telephone: +43 1 2600 22529 or 22530
Email: sales.publications@iaea.org
www.iaea.org/publications

Priced and unpriced IAEA publications may be ordered directly from the IAEA.

ORDERING LOCALLY

Priced IAEA publications may be purchased from regional distributors and from major local booksellers.

Printed and bound by CPI Group (UK) Ltd, Croydon, CR0 4YY

06/07/2026

02160600-0004